Advances and Technologies in Building Construction and Structural Analysis

Edited by Alireza Kaboli and Sara Shirowzhan

Published in London, United Kingdom

IntechOpen

Supporting open minds since 2005

Advances and Technologies in Building Construction and Structural Analysis
http://dx.doi.org/10.5772/intechopen.88410
Edited by Alireza Kaboli and Sara Shirowzhan

Contributors
Alonso Gomez-Bernal, Eduardo Arellano-Mendez, Hugon Juarez-García, Angel Quiroz, Oscar González-Cuevas, Wengui Li, Bin Lei, Zhiyu Luo, Fuzhi Yang, Zhuo Tang, Aman Kumar, Jasvir Singh Rattan, Ajay Kumar, Nishant Raj Kapoor, Rahul Kumar, Anqi Shi, Sara Shirowzhan, Samad M.E. M.E. Sepasgozar, Takao Katsura, Reyhaneh Karimi, Sharifeh Sargolzaei, Leila Farahzadi, Akram Nasrolahi, Sanee M. Ebrahimzadeh Sepasgozar, Mohsen Zareian, Rafaat Hussein

First published in London, United Kingdom, 2021 by IntechOpen
IntechOpen is the global imprint of INTECHOPEN LIMITED, registered in England and Wales, registration number: 11086078, 5 Princes Gate Court, London, SW7 2QJ, United Kingdom
Printed in Croatia

British Library Cataloguing-in-Publication Data
A catalogue record for this book is available from the British Library

Additional hard and PDF copies can be obtained from orders@intechopen.com

Advances and Technologies in Building Construction and Structural Analysis
Edited by Alireza Kaboli and Sara Shirowzhan
p. cm.
Print ISBN 978-1-83881-140-2
Online ISBN 978-1-83881-141-9
eBook (PDF) ISBN 978-1-83881-142-6

An electronic version of this book is freely available, thanks to the support of libraries working with Knowledge Unlatched. KU is a collaborative initiative designed to make high quality books Open Access for the public good. More information about the initiative and links to the Open Access version can be found at www.knowledgeunlatched.org

For EU product safety concerns:
IN TECH d.o.o., Prolaz Marije Krucifikse Kozulić 3, 51000 Rijeka, Croatia, info@intechopen.com or visit our website at intechopen.com.

Meet the editors

Dr. Alireza Kaboli obtained an MEng in Structural Engineering from Semnan University, Iran, in 2007, and a Ph.D. in Construction Engineering and Management from the University of New South Wales (UNSW) Sydney, Australia, in 2014. He previously served as a lecturer at Azad University, Iran, and a guest lecturer at UNSW. Dr. Kaboli has published more than twenty conference and journal papers.

Dr. Sara Shirowzhan is a lecturer and Academic Advisor for the City Analytics Program, School of Built Environment, University of New South Wales (UNSW), Sydney, Australia. Her areas of research include sensing technologies, advanced GIS, BIM, digital twins, and artificial intelligence. She teaches and supervises students on city analytics, GIS, construction informatics, and BIM topics at undergraduate and postgraduate levels at UNSW. Dr. Shirowzhan is currently an editorial board member for the journal Sustainability and Advances in Civil Engineering. She is also a topic board member of the ISPRS International Journal of Geo-Information. Dr. Shirowzhan obtained her Ph.D. in Geomatics Engineering from the School of Civil and Environmental Engineering, UNSW.

Contents

Preface

This Edited Volume is a collection of reviewed and relevant research chapters, concerning the developments within the "Advances and Technologies in Building Construction and Structural Analysis" field of study. The book includes scholarly contributions by various authors and is edited by a group of experts pertinent to alternative medicine. Each contribution comes as a separate chapter complete in itself but directly related to the book's topics and objectives.

The book has 8 chapters. The chapters are in the following order: "Treatise on Sustainable Infrastructure Construction: Green Composites, Cross Laminated/Mass Timber, Wood Truss Connectors, Nondestructive Technologies, Health Assessment and Monitoring: Utility Poles and Geofoam", "Construction Technology of Precast Pier Foundation Filled with Demolished Concrete Lumps", "Vertical Bearing Capacity of Precast Pier Foundation Filled with Demolished Concrete Lumps", "Structural Health Monitoring of Existing Reinforced Cement Concrete Buildings and Bridge Using Nondestructive Evaluation with Repair Methodology", "Behavior and Design of Transfer Slabs Subjected to Shear Wall Loads", "Computational Workflow for Three-Dimension Printing in Construction: Digital Tools and Methodological Limitations", "Transparent Vacuum Insulation Panels" and "Smart Built Environment Including Smart Home, Smart Building and Smart City: Definitions and Applied Technologies."

The target audience comprises scholars and specialists in the field.

IntechOpen

Chapter 1

Treatise on Sustainable Infrastructure Construction: Green Composites, Cross Laminated/Mass Timber, Wood Truss Connectors, Nondestructive Technologies, Health Assessment and Monitoring: Utility Poles and Geofoam

Rafaat Hussein

Abstract

The understanding of the engineering performance of green laminated composites is necessary to the design of load bearing components in building and infrastructure construction, and packaging applications. These components are made of outer thin laminae called skins or faces and a thick inner layer called core. The use of bonding is unavoidable in the assembling of these composite products. Like all materials, the bonding materials have finite mechanical properties, e.g. stiffness, but when used in the literature, they are assumed perfectly rigid. That is an unrealistic assumption. Our analytical solutions change this assumption by using the real properties of bonding. In general, the analytical formulations are based on the equilibrium equations of forces, the compatibility of interlaminar stresses and deformation, and the geometrical conditions of the panels. Once solutions are obtained, the next step is to evaluate them. The numerical evaluations proved that perfect rigid bonding in laminated composites greatly underestimates the true performance. At low values of adhesive stiffness, the serviceability is multiple orders of magnitude of that at high values. The logical question is thus: what constitutes perfect bonding? The answer to this question lies in the core-to-adhesive stiffness. The lower the ration is the higher the error in using the rigid-bond theories. It is worth noting that green-composites in this chapter refer to components made of traditional materials such as wood, in addition to newly developed bio-based and bio-degradable and bio-based composites, made of renewable resources. In addition, bonding and adhesive are used interchangeably.

Keywords: bonding, composites, laminates, rigid-bonding, slip, stiffness

1. Overall introduction

From societal point of view, we must recognize that infrastructure components, such as buildings, represent the physical systems that provide a continual flow of the essential daily services in any society. Their disrepair and degradation disrupt the functioning of business, government, and facilities, as well as have cascading effects throughout the world. Report after report reinforces this prevailing ailing situation. The American Society of Civil Engineers assesses the national status of the infrastructure about each four years. The overall grade point average since 2009 has been around D grade. This is a sign of the needs for urgent actions to repair or replace the ailing systems. The basic challenge is that we must find ways to design new buildings with long life spans so that the annual maintenance required is much

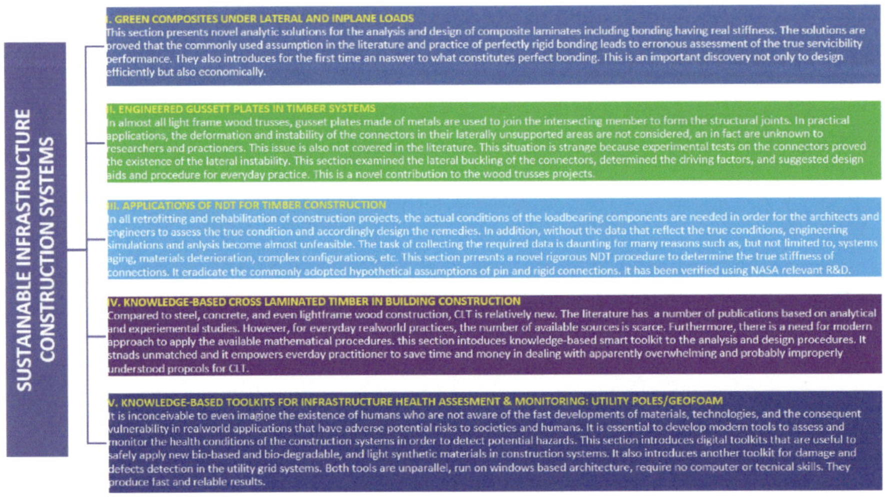

Figure 1.
Overall chapter organization.

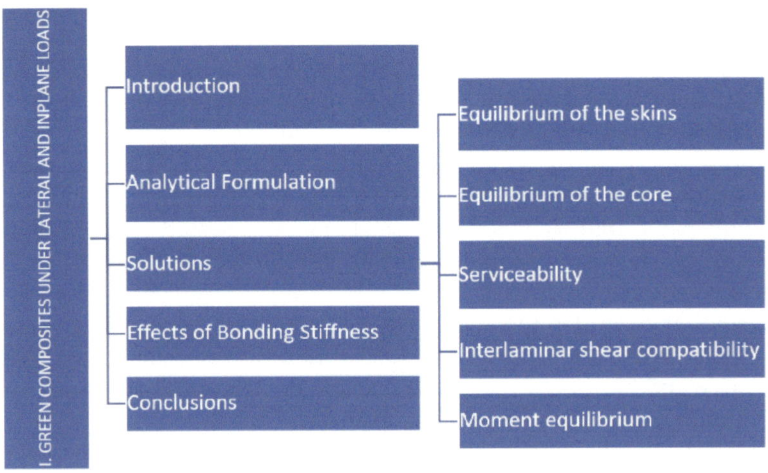

Figure 2.
Part I: Green composites.

less than at present. This chapter is a modern treatise on five practical advances that meet some of the challenges as shown in **Figures 1–6**. Taken as a whole, the chapter brings complex or unsolved problems to everyday practice.

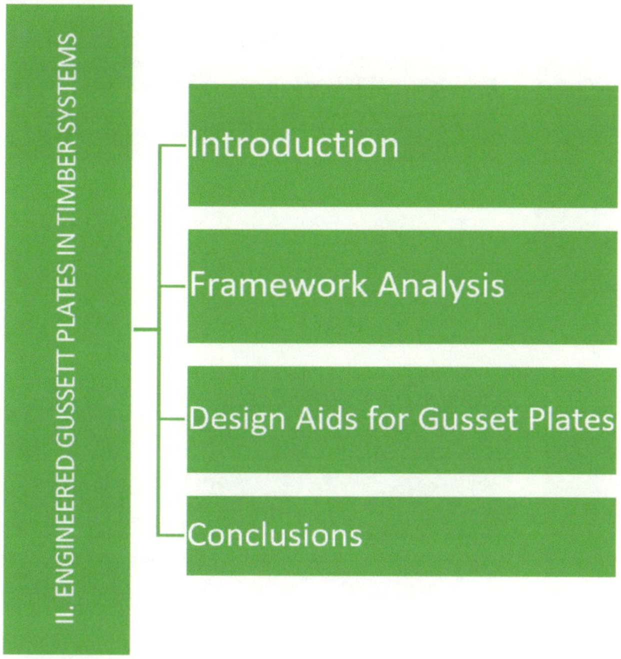

Figure 3.
Engineered Gussett plates in timber systems.

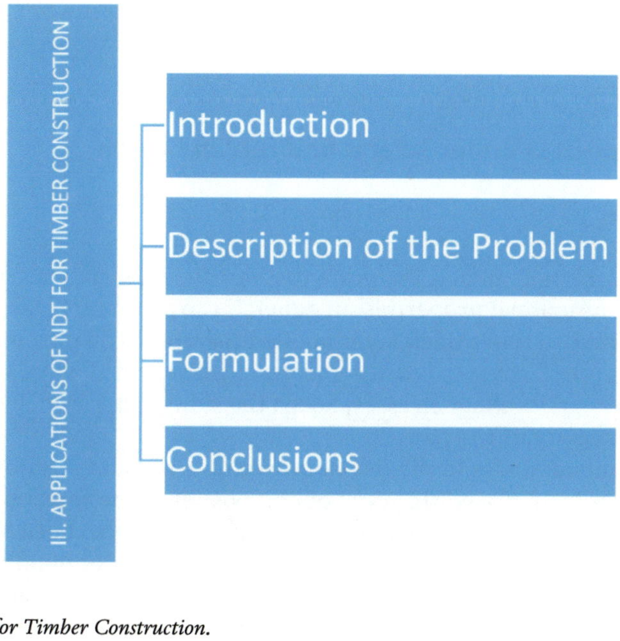

Figure 4.
NDT Technologies for Timber Construction.

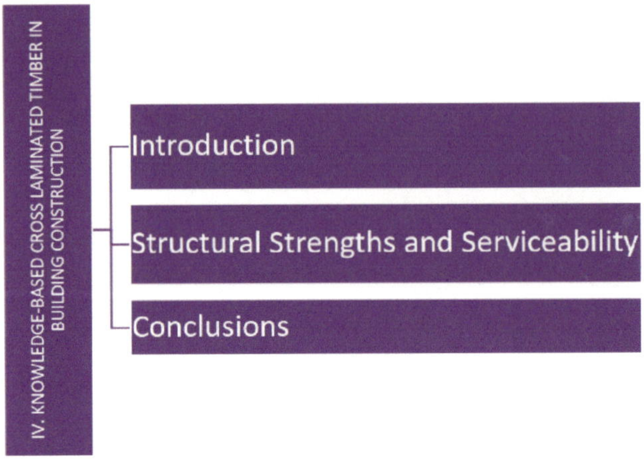

Figure 5.
KB toolkit for CLT.

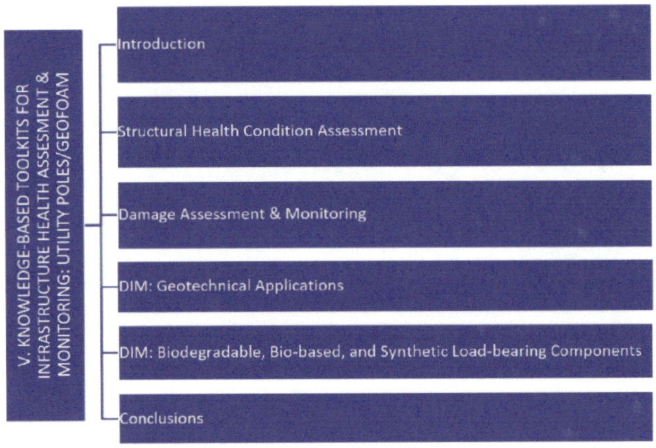

Figure 6.
KB toolkits for utility poles and Geofoam.

2. Green composites under lateral and inplane loads

2.1 Introduction

Laminated composite panels integrate the benefits of each lamina including mechanical and functional properties. In general, the panels are assembled of outer stiff thin layers and an inner thick layer bonded together. The purpose of the face or skin is to carry normal stresses resulting from in-plane and flexural deformations, whereas the core carries shear stresses, keeps the skins apart, and is responsible for the structural integrity of the panel.

The significant advances in the development of green and environmentally friendly materials have placed this type of composites at the forefront of research and development [1–9]. Although laminated composites have been used for over a century, the literature lacks adequate studies on the effects of bonding stiffness on the overall performance. To narrow the gap of the missing information, the author has completed some analytic and experimental studies [4] in this area. This chapter takes the previous studies a step farther. It ascertains the effects of using bonding materials with finite mechanical properties on the response of the load bearing components in construction. Lateral flexural loads, such as gravity, wind, and pressures, are inevitable in almost all buildings, and packaging applications. In addition, the use of adhesives is also unavoidable for the manufacturing of any laminated composites. The logical question is thus: what affects the realistic adhesive properties, such as stiffness, have on the flexural response of the load bearing panels? The answer to this question provides new knowledge on the true aspects of the structural response and the driving factors. It also narrows the gap in the published articles that assumed perfect rigid bonding layers in laminated composites. Further studies are needed to include damage formation and propagation, localized deformation, and nonlinearity [10–12].

2.2 Analytical formulation

Existing analytic and experimental methods have invariably assumed perfectly rigid bonding between layers [8, 13–15]. Nevertheless, interlaminar delamination occur because of the finite bonding stiffness; the bonds creep under sustained loads, and environmental effects. The high interlaminar shear stress due to applied loads may contribute to an answer of the many delamination in building and infrastructure construction applications. Analysis of wood joist floor systems, commonly used in buildings, including interlaminar shear deformations is available in the literature [8, 16–18]. The wood layers were assembled with nails with glue at their ends. Although the interlaminar deformation in this system was accounted for in the analytical model, transverse shear deformations were neglected. Very few publications deal with the responses of laminated composites with interlaminar deformation or orthotropic material properties. To ascertain the effects of the unrealistic assumption of perfect rigid bonding on the serviceability and strengths of load bearing components in construction, the author has conducted a series of analytical and experimental studies on laminated composites under various loads [14, 19]. The studies used realistic mechanical properties for the bonding materials, introduced in the analytical formulation new parameters such as the core properties, and the in-plane and inter-laminar deformation.

This chapter presents a closed-form solution of laminated composites with interlaminar deformations and under flexural loads. The solution satisfies the equilibrium equations of each lamina and the compatibility of deformations at the interfaces. One of the objectives is to ascertain the effect of interlaminar deformations on the performance of laminates due to transverse flexural loads which are everyday loads in building and infrastructure construction.

2.3 Solutions

The analytic investigation of the laminates at hands is sufficiently mathematically complex due to the existence of non-zero shear and normal strains in the core and bonding. Consider a panel of span 2a and width 2b, as shown in **Figure 7**. The faces are of equal thin thickness t_f. The core of a thickness t_c. The moduli of elasticity of the skins and core are E_f and E_c, respectively. The shear modulus of the core is G_c. The bonding between the laminae has finite stiffness K.

By using the fundamentals of theory of elasticity [19–23], equations are set up to define the equilibrium of the skins and of core elements, and to prescribe the strains continuity between them. The result is a set of differential equations which one solved together lead to the quantities that characterize the overall performance. The stress state in the faces and core elements is shown in **Figure 8**. The equilibrium of the face element requires that

$$\frac{\partial \sigma^{j}_{fx}}{\partial x} + \frac{\partial \tau^{j}_{yx}}{\partial y} + \frac{q_x}{t^{j}_{f}} = 0 \tag{1}$$

$$\frac{\partial \sigma^{j}_{fy}}{\partial y} + \frac{\partial \tau^{j}_{xy}}{\partial x} + \frac{q_y}{t^{j}_{f}} = 0 \tag{2}$$

in which σ_{fx}, σ_{fy}, τ_{xy} = normal and shear stress components in the skins; q_i = interlaminar shear stress in the i direction (i = x or y); j = superscript denoting stresses in the face j (j = 1 bottom skin), x, y = coordinate axes.

The state of stresses in the core must satisfy the following equilibrium equations

$$\frac{\partial \sigma_{cx}}{\partial x} + \frac{\partial \tau_{czx}}{\partial z} = 0 \tag{3}$$

$$\frac{\partial \sigma_{cy}}{\partial y} + \frac{\partial \tau_{czy}}{\partial z} = 0 \tag{4}$$

in which σ_{ci}, = normal stress in the skin in the i direction, τ_{czi} = shear stress components in the core. Eqs. 3 and 4 can be written in terms of the displacement components

$$E_{cx}\frac{\partial^2 u_c}{\partial x^2} + G_{cx}\frac{\partial^2 u_c}{\partial z^2} = 0 \tag{5}$$

$$E_{cy}\frac{\partial^2 \vartheta_c}{\partial y^2} + G_{cy}\frac{\partial^2 \vartheta_c}{\partial z^2} = 0 \tag{6}$$

in which u_c and υ_c = displacement in the core along the x and y directions, respectively; E and G = the elastic and shear modulii of the core material.

The stress components in the faces and core must also satisfy the equilibrium equation in terms of the applied load, which is

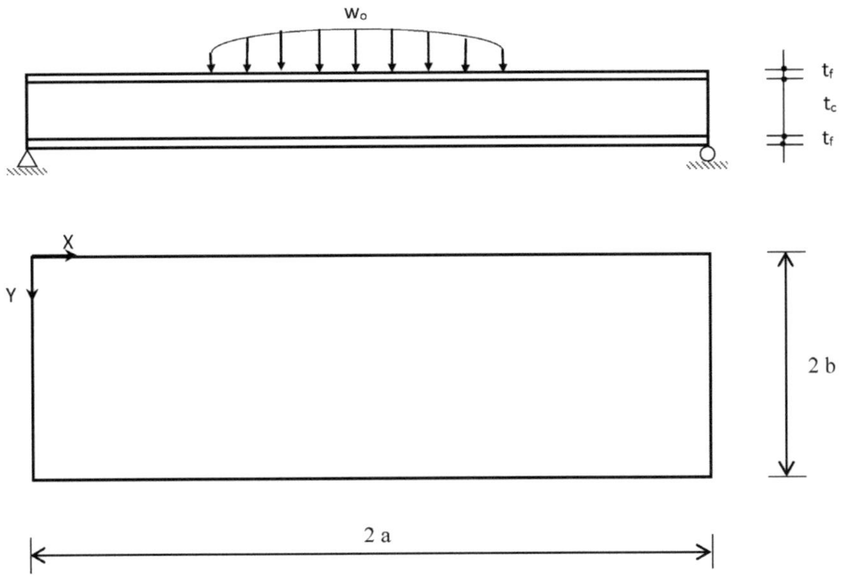

Figure 7.
A typical laminate under transverse load.

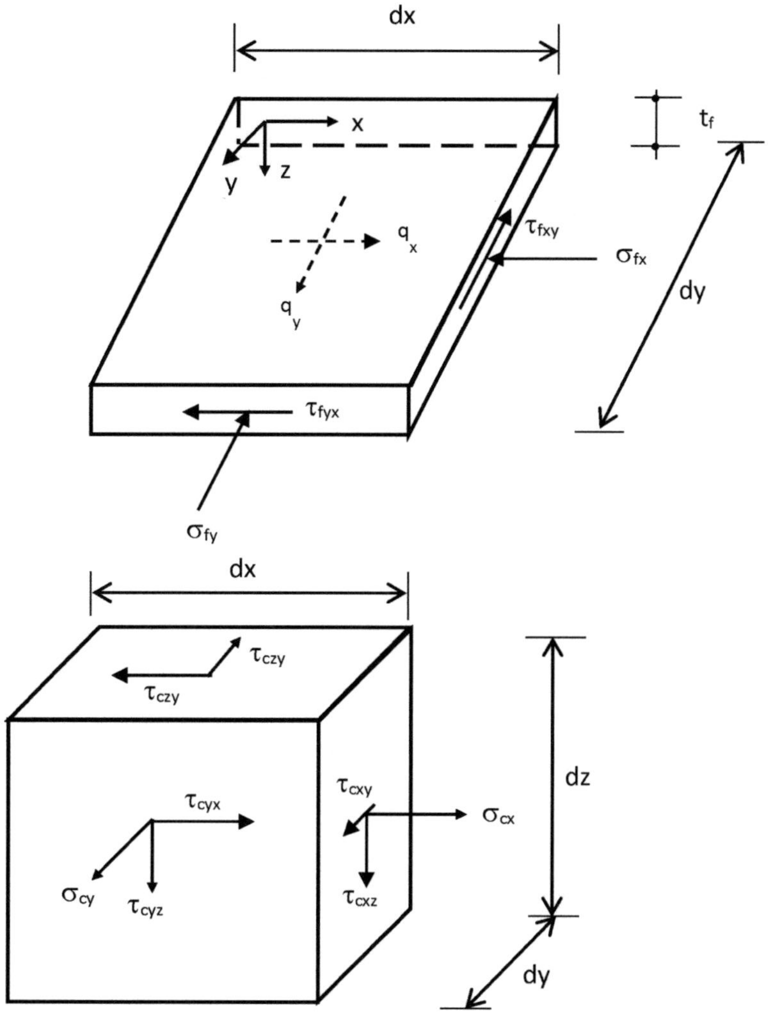

Figure 8.
Stress state in laminates' elements.

$$\frac{\partial^2 M_x}{\partial x^2} - 2\frac{\partial^2 M_{xy}}{\partial x\,\partial y} + \frac{\partial^2 M_y}{\partial y^2} = -p \tag{7}$$

in which M_x, M_y; and M_{xy} = the applied moments; and p = the applied load intensity.

At the interfaces between the core and the skins, the stresses and strains must be compatible. The compatibility equations in terms of stresses are

$$q_x = G_{cx}\left(\frac{\partial u_c}{\partial z} + \frac{\partial \omega}{\partial x}\right)_{z=\frac{1}{2}t_c} \tag{8}$$

$$q_y = G_{cy}\left(\frac{\partial \vartheta_c}{\partial z} + \frac{\partial \omega}{\partial y}\right)_{z=\frac{1}{2}t_c} \tag{9}$$

in which t_c = the core thickness; and ω = the lateral deflection of the laminates. In terms of strains, the compatibility equations are written as

$$\frac{\partial \Delta_x}{\partial x} = \varepsilon_{fx} - (\varepsilon_{cx})_{z=\frac{1}{2}t_c} \tag{10}$$

$$\frac{\partial \Delta_y}{\partial y} = \varepsilon_{fy} - (\varepsilon_{cy})_{z=\frac{1}{2}t_c} \tag{11}$$

in which.
ε = normal strain = ∂u/∂x or ∂υ/∂y.
i = x or y direction.
f, c = face and core, respectively.
Δ = interlayer shearing slip = q/K.
K = adhesive stiffness.
In addition to satisfying the force and stress conditions, a solution must also satisfy the geometrical conditions. With respect to the laminates shown in **Figure** 7, the relevant boundary conditions are

$$\sigma_{fx} = 0, \sigma_{cx} = \gamma_y = 0, \text{and } \omega = 0 \text{ at } x = 0 \text{ and } x = 2a \tag{12}$$

$$\sigma_{fy} = 0, \sigma_{cy} = \gamma_x = 0, \text{and } \omega = 0 \text{ at } x = 0 \text{ and } x = 2b \tag{13}$$

$$u_c = \upsilon_c = 0 \text{ at the middle plane of the laminates, i.e.} z = 0 \tag{14}$$

$$u_f = u_c = 0 \text{ at } x = a \text{ and } \upsilon_f = \upsilon_c = 0 \text{ at } y = b \text{ in laminates with symmetrical loads} \tag{15}$$

in which 2a and 2b = the laminates dimensions in x and y directions, respectively; γ_i = the shear strain in the z_i plane, and u_f and υ_f = the displacements in the skin along the x and y directions, respectively.

2.4 Equilibrium of the skins

For the laminates in **Figure** 7, the displacement components in the faces are considered as

$$u_f = \sum_{m=1,3,\ldots}^{\infty} \sum_{n=1,3,\ldots}^{\infty} A_{mn} \cos \alpha_m x \sin\beta_n \tag{16}$$

$$\vartheta_f = \sum_{m=1,3,\ldots}^{\infty} \sum_{n=1,3,\ldots}^{\infty} B_{mn} \sin \alpha_m x \cos\beta_n y \tag{17}$$

in which α_m = m π/2a, β_n = n π/2b, and A_{mn} and B_{mn} are unknown coefficients. From Eqs. 16 and 17, expressions for the stresses in the faces can be derived as

$$\sigma_{fx} = \frac{E_f}{\nu^2 - 1} \sum_{m=1,3,\ldots}^{\infty} \sum_{n=1,3,\ldots}^{\infty} (A_{mn}\alpha_m + \nu B_{mn}\beta_n) \sin \alpha_m x \sin \beta_n y \tag{18}$$

$$\sigma_{fy} = \frac{E_f}{\nu^2 - 1} \sum_{m=1,3,\ldots}^{\infty} \sum_{n=1,3,\ldots}^{\infty} (B_{mn}\beta_n + \nu A_{mn}\alpha_m) \sin \alpha_m x \sin \beta_n y \tag{19}$$

$$\tau_{xy} = \frac{E_f}{2(1+\nu)} \sum_{m=1,3,\ldots}^{\infty} \sum_{n=1,3,\ldots}^{\infty} (A_{mn}\beta_n + B_{mn}\alpha_m) \cos \alpha_m x \cos \beta_n y \tag{20}$$

Expressions for the interlaminar shear stresses q_x and q_y are obtained by differentiating Eqs. 18 to 20 in accordance with the equilibrium Eqs. 1 and 2

$$q_x = \frac{E_f t_f}{1+\nu} \sum_{m=1,3,\dots}^{\infty} \sum_{n=1,3,\dots}^{\infty} \left(\frac{A_{mn}\alpha_m^2 + \nu B_{mn}\vartheta_n\alpha_m}{\nu - 1} - \frac{A_{mn}\beta_n^2 + B_{mn}\alpha_m\beta_n}{2} \right) \cos\alpha_m x \sin\beta_n y \tag{21}$$

$$q_y = \frac{E_f t_f}{1+\nu} \sum_{m=1,3,\dots}^{\infty} \sum_{n=1,3,\dots}^{\infty} \left(\frac{B_{mn}\beta_n^2 + \nu A_{mn}\alpha_m\beta_n}{\nu - 1} - \frac{B_{mn}\alpha_m^2 + A_{mn}\alpha_m\beta_n}{2} \right) \sin\alpha_m x \cos\beta_n y \tag{22}$$

2.5 Equilibrium of the core

A solution for the core displacement satisfying the equilibrium Eqs. 5 and 6 as well as the boundary conditions in Eqs. 12–15 is taken as

$$u_c = \sum_{m=1,3,\dots}^{\infty} C_m \cos\,\alpha_m x\ \sinh\,\mu_x\alpha_m z \tag{23}$$

$$\vartheta_c = \sum_{m=1,3,\dots}^{\infty} D_n \cos\,\beta_n y\ \sinh\,\mu_y\beta_n z \tag{24}$$

in which $\mu_x = (E_{cx}/G_{cx})^{1/2}$; $\mu_y = (E_{cy}/G_{cy})^{1/2}$; and C_m and D_n are unknown functions of y and x, respectively.

2.6 Serviceability

For the laminates in **Figure 7**, the deflection function which satisfies the boundary conditions in Eq. 12 is given by

$$\omega = \sum_{m=1,3,\dots}^{\infty} W_m\ \sin\,\alpha_m x \sin\,\beta_n y \tag{25}$$

in which W_m = unknown coefficient.

2.7 Interlaminar shear compatibility

The functions C_m and D_n can now be expressed in terms of A_{mn} and B_{mn} by using the compatibility Eqs. 10 and 11. By substituting Eqs. 16, 21, and 23 in Eq. 10 and expression for C_m is obtained as

$$C_m = \sum_{n=1,3,\dots}^{\infty} \frac{1}{\sinh\left(\frac{\mu_x\alpha_m t_c}{2}\right)} \left[\frac{E_f t_f}{K_x(1+\nu)} \left(\frac{A_{mn}\alpha_m^2 + \nu B_{mn}\beta_n\alpha_m}{1-\nu} + \frac{A_{mn}\beta_n^2 + B_{mn}\alpha_m\beta_n}{2} \right) + A_{mn} \right] \sin\,\beta_n y \tag{26}$$

In a similar manner, D_n is obtained from Eqs. 17, 22, 24, and 11 as

$$D_n = \sum_{m=1,3,\dots}^{\infty} \frac{1}{\sinh\left(\frac{\mu_y\beta_n t_c}{2}\right)} \left[\frac{E_f t_f}{K_y(1+\nu)} \left(\frac{B_{mn}\beta_n^2 + \nu A_{mn}\beta_n\alpha_m}{1-\nu} + \frac{B_{mn}\alpha_n^2 + A_{mn}\alpha_m\beta_n}{2} \right) + A_{mn} \right] \sin\,\beta_n y \tag{27}$$

The coefficient B_{mn} can also be expressed in terms of A_{mn} by using the compatibility Eq. 8. By substituting ω, q_x, and u_c in the first equation of Eq. 8, and by equating the coefficients of $\cos\,\alpha_m$ x $\sin\,\beta_n$ y in both of its sides, it is found that

$$W_m = \frac{E_f t_f}{G_{cx}(1+\nu)}\left(\frac{A_{mn}\alpha_m + \nu\, B_{mn}\beta_n}{\nu - 1} - \frac{A_{mn}\beta_n^2 + B_{mn}\alpha_m\beta_n}{2\,\alpha_m}\right) - \mu_x \coth\left(\frac{\mu_x\alpha_m t_c}{2}\right)$$
$$\times\left[\frac{E_f t_f}{K_x(1+\nu)}\left(\frac{A_{mn}\alpha_m^2 + \nu\, B_{mn}\beta_n\alpha_m}{1-\nu} + \frac{A_{mn}\beta_n^2 + B_{mn}\alpha_m\beta_n}{2}\right) + A_{mn}\right] \tag{28}$$

In a similar manner, from Eqs. 22, 24, and the second equation in Eq. 9, W_m is found as

$$W_m = \frac{E_f t_f}{G_{cy}(1+\nu)}\left(\frac{B_{mn}\beta_n + \nu\, A_{mn}\alpha_m}{\nu - 1} - \frac{B_{mn}\alpha_n^2 + A_{mn}\alpha_m\beta_n}{2\,\beta_n}\right) - \mu_y \coth\left(\frac{\mu_y\beta_n t_c}{2}\right)$$
$$\times\left[\frac{E_f t_f}{K_y(1+\nu)}\left(\frac{B_{mn}\beta_n^2 + \nu\, B_{mn}\beta_n\alpha_m}{1-\nu} + \frac{B_{mn}\alpha_n^2 + A_{mn}\alpha_m\beta_n}{2}\right) + B_{mn}\right] \tag{29}$$

By equating the right-hand sides of Eqs. 28 and 29, it follows that

$$B_{mn} = \frac{\phi_{mn}}{\psi_{mn}} A_{mn} \tag{30}$$

In which

$$\phi_{mn} = \frac{E_f t_f}{G_{cy}(1+\nu)}\left[-\frac{\alpha_m}{2} + \frac{G_{cy}}{2\,G_{cx}}\frac{\beta_n^2}{\alpha_m} + \frac{G_{cy}\mu_y}{2\,K_y}\frac{\nu+1}{\nu-1}\alpha_m\beta_n \coth\left(\frac{\mu_y\beta_n t_c}{2}\right)\right.$$
$$+\frac{G_{cy}\mu_x}{2\,K_x}\left(\frac{2\,\alpha_m^2}{1-\nu} + \beta_n^2\right)\coth\left(\frac{\mu_x\alpha_m t_c}{2}\right) + \frac{G_{cy}}{G_{cx}}\frac{\alpha_m}{1-\nu} + \frac{\nu\,\alpha_m}{\nu-1}$$
$$\left. +\mu_x \frac{G_{cy}(\nu+1)}{E_f t_f}\coth\left(\frac{\mu_x\alpha_m t_c}{2}\right)\right] \tag{31}$$

$$\psi_{mn} = \frac{E_f t_f}{G_{cx}(1+\nu)}\left[-\frac{\beta_n}{2} + \frac{G_{cx}}{2\,G_{cy}}\frac{\alpha_m^2}{\beta_n} + \frac{G_{cx}\mu_x}{2\,K_x}\frac{\nu+1}{\nu-1}\alpha_m\beta_n \coth\left(\frac{\mu_x\alpha_m t_c}{2}\right)\right.$$
$$+\frac{G_{cx}\mu_y}{2\,K_y}\left(\frac{2\,\beta_n^2}{1-\nu} + \alpha_m^2\right)\coth\left(\frac{\mu_y\beta_n t_c}{2}\right) + \frac{G_{cx}}{G_{cy}}\frac{\beta_n}{1-\nu} + \frac{\nu\,\beta_n}{\nu-1}$$
$$\left. +\mu_y \frac{G_{cx}(\nu+1)}{E_f t_f}\coth\left(\frac{\mu_y\beta_n t_c}{2}\right)\right] \tag{32}$$

2.8 Moment equilibrium

The only unknown coefficient now is A_{mn}. This can be determined by using the equilibrium Eq. 7. The moments M_x, M_y, and M_{xy} can be expressed in terms of the stresses components as

$$M_x = \sigma_{fx} t_f\, h + \int_{-\frac{1}{2}t_f}^{+\frac{1}{2}t_f} \sigma_{cx}\, z\, dz \tag{33}$$

$$M_y = \sigma_{fy} t_f\, h + \int_{-\frac{1}{2}t_f}^{+\frac{1}{2}t_f} \sigma_{cy}\, z\, dz \tag{34}$$

$$M_{xy} = -\tau_{xy} t_f\, h \tag{35}$$

in which $h = t_c + t_f$; $\sigma_{cx} = E_{cx}\,(\partial u_c/\partial x)$; and $\sigma_{cy} = E_{cy}\,(\partial \upsilon_c/\partial y)$. In terms of these stresses, the equilibrium Eq. 7 becomes

$$t_f\,h\left(\frac{\partial^2\sigma_{fx}}{\partial x^2}+2\,\frac{\partial^2\tau_{xy}}{\partial x\,\partial y}+\frac{\partial^2\sigma_{fy}}{\partial y^2}\right)+E_{cx}\frac{\partial^2}{\partial x^2}\int_{-\frac{1}{2}t_f}^{+\frac{1}{2}t_f}\frac{\partial u_c}{\partial x}\,z\,dz+E_{cy}\frac{\partial^2}{\partial y^2}\int_{-\frac{1}{2}t_f}^{+\frac{1}{2}t_f}\frac{\partial \vartheta_c}{\partial y}\,z\,dz=-p \tag{36}$$

Substituting Eqs. 18–20, 23, and 24 in Eq. 35 yields the following expression for A_{mn}

$$A_{mn}=\frac{P_{mn}}{\xi_{mn}} \tag{37}$$

in which P_{mn} = Fourier series expansion of the applied load, p.

$$\begin{aligned}\xi_{mn}=&\frac{E_f t_f\,h}{\nu^2-1}\left(\alpha_m^3+\nu\,\beta_n\alpha_m^2\frac{\phi_{mn}}{\psi_{mn}}\right)-\frac{E_f t_f\,h}{1+\nu}\left(\alpha_m\beta_n^2+\beta_n\alpha_m^2\frac{\phi_{mn}}{\psi_{mn}}\right)\\&+\frac{E_f t_f\,h}{\nu^2-1}\left(\nu\,\alpha_m\beta_n^2+\beta_n^3\frac{\phi_{mn}}{\psi_{mn}}\right)\\&+\frac{2\,E_{cx}\mu_y C'_{mn}}{\mu_x^2}\left(\frac{\mu_x\alpha_m t_c}{2}\cosh\left(\frac{\mu_x\alpha_m t_c}{2}\right)-\sinh\left(\frac{\mu_x\alpha_m t_c}{2}\right)\right)\\&+\frac{2\,E_{cy}\beta_n D'_{mn}}{\mu_y^2}\left(\frac{\mu_y\beta_n t_c}{2}\cosh\left(\frac{\mu_y\beta_n t_c}{2}\right)-\sinh\left(\frac{\mu_y\beta_n t_c}{2}\right)\right)\end{aligned} \tag{38}$$

in which

$$C'_{mn}=\frac{1}{\sinh\left(\frac{\mu_x\alpha_m t_c}{2}\right)}\left[\frac{E_f t_f}{K_x(1+\nu)}\left(\frac{\alpha_m^2+\nu\,\beta_n\alpha_m\frac{\phi_{mn}}{\psi_{mn}}}{\nu-1}-\frac{\beta_n^2+\beta_n\alpha_m\frac{\phi_{mn}}{\psi_{mn}}}{2}\right)-1\right] \tag{39}$$

$$D'_{mn}=\frac{1}{\sinh\left(\frac{\mu_y\beta_n t_c}{2}\right)}\left[\frac{E_f t_f}{K_y(1+\nu)}\left(\frac{\nu\,\beta_n\alpha_m+\beta_n^2\frac{\phi_{mn}}{\psi_{mn}}}{\nu-1}-\frac{\beta_n\alpha_m+\alpha_m^2\frac{\phi_{mn}}{\psi_{mn}}}{2}\right)-\frac{\phi_{mn}}{\psi_{mn}}\right] \tag{40}$$

In the particular case where $G_{cx} = G_{cy} = G_c$; $E_{cx} = E_{cy} = E_c$; and $K_x = K_y = K$, hence $\mu_x = \mu_y = (E_c/G_c)^{1/2} = \mu$, thus the expressions for ϕ_{mn}, ψ_{mn}, and ξ_{mn} become

$$\begin{aligned}\xi_{mn}=&\frac{E_f t_f h}{G_c(1+\nu)}\left[\frac{\alpha_m}{2}+\frac{\beta_n^2}{2\,\alpha_m}+\frac{G_c\mu\,\beta_n\alpha_m}{2\,K}\frac{\nu+1}{\nu-1}\coth\left(\frac{\mu\,\beta_n t_c}{2}\right)\right.\\&\left.+\frac{G_c\mu}{2\,K}\left(\frac{2\,\alpha_m^2}{1-\nu}+\beta_n^2\right)\coth\left(\frac{\mu\,\alpha_m t_c}{2}\right)+\frac{G_c\,\mu\,(1+\nu)}{E_f t_f}\coth\left(\frac{\mu\,\alpha_m t_c}{2}\right)\right]\end{aligned} \tag{41}$$

$$\begin{aligned}\psi_{mn}=&\frac{E_f t_f}{G_c(1+\nu)}\left[\frac{\beta_n}{2}+\frac{\alpha_m^2}{2\,\beta_n}+\frac{G_c\mu\,\beta_n\alpha_m}{2\,K}\frac{\nu+1}{\nu-1}\coth\left(\frac{\mu\,\alpha_m t_c}{2}\right)\right.\\&\left.+\frac{G_c\mu}{2\,K}\left(\frac{2\,\beta_n^2}{1-\nu}+\alpha_m^2\right)\coth\left(\frac{\mu\,\beta_n t_c}{2}\right)+\frac{G_c\,\mu\,(1+\nu)}{E_f t_f}\coth\left(\frac{\mu\,\beta_n t_c}{2}\right)\right]\end{aligned} \tag{42}$$

$$\begin{aligned}\xi_{mn}=&\frac{E_f t_f\,h}{\nu^2-1}\left(\alpha_m^3+\nu\,\beta_n\alpha_m^2\frac{\phi_{mn}}{\psi_{mn}}\right)-\frac{E_f t_f\,h}{\nu+1}\left(\alpha_m\,\beta_n^2+\beta_n\alpha_m^2\frac{\phi_{mn}}{\psi_{mn}}\right)\\&+\frac{E_f t_f\,h}{\nu^2-1}\left(\nu\,\alpha_m\,\beta_n^2+\beta_n^3\frac{\phi_{mn}}{\psi_{mn}}\right)\\&+2\,G_c\alpha_m C'_m\left(\frac{\mu\,\alpha_m t_c}{2}\cosh\left(\frac{\mu\,\alpha_m t_c}{2}\right)-\sinh\left(\frac{\mu\,\alpha_m t_c}{2}\right)\right)\\&+2\,G_c\beta_n D'_n\left(\frac{\mu\,\beta_n t_c}{2}\cosh\left(\frac{\mu\,\beta_n t_c}{2}\right)-\sinh\left(\frac{\mu\,\beta_n t_c}{2}\right)\right)\end{aligned} \tag{43}$$

in which

$$C'_{mn} = \frac{1}{\sinh\left(\frac{\mu\,\alpha_m t_c}{2}\right)}\left[\frac{E_f t_f}{K(1+\nu)}\left(\frac{\alpha_m^2 + \nu\,\beta_n\alpha_m\frac{\phi_{mn}}{\psi_{mn}}}{\nu-1} - \frac{\beta_n^2 + \beta_n\alpha_m\frac{\phi_{mn}}{\psi_{mn}}}{2}\right) - 1\right] \quad (44)$$

$$D'_{mn} = \frac{1}{\sinh\left(\frac{\mu\,\beta_n t_c}{2}\right)}\left[\frac{E_f t_f}{K(1+\nu)}\left(\frac{\nu\,\beta_n\alpha_m + \beta_n^2\frac{\phi_{mn}}{\psi_{mn}}}{\nu-1} - \frac{\beta_n\alpha_m + \alpha_m^2\frac{\phi_{mn}}{\psi_{mn}}}{2}\right) - \frac{\phi_{mn}}{\psi_{mn}}\right] \quad (45)$$

By following the above analytical formulation and considering the in-plane load shown in **Figure 9**, the following solution is obtained

$$\sigma_{fx} = \sum_{m=1,3,..}^{\infty}\sum_{n=1,3,..}^{\infty}\left[A_{mn}\lambda'_{z1} + C_{mn}\lambda'_{z2}\right]S_x S_y + \left(\sigma_{xo} + \sigma_{cxo}\frac{t_c}{t_f}\right)\sum_{m=1}^{\infty}\sum_{n=1}^{\infty}\frac{2}{a\,\alpha_m}\frac{2}{b\,\beta_n}S_x S_y \quad (46)$$

$$\sigma_{fy} = \sum_{m=1,3,..}^{\infty}\sum_{n=1,3,..}^{\infty}\left[A_{mn}\lambda'_{z3} + C_{mn}\lambda'_{z4}\right]S_x S_y + \left(\sigma_{yo} + \sigma_{cyo}\frac{t_c}{t_f}\right)\sum_{m=1}^{\infty}\sum_{n=1}^{\infty}\frac{2}{a\,\alpha_m}\frac{2}{b\,\beta_n}S_x S_y \quad (47)$$

$$\tau_{fxy} = \sum_{m=1,3,..}^{\infty}\sum_{n=1,3,..}^{\infty}A_{mn}\left[\frac{G_{fxy}}{E_{cx}}\phi_x\Big|_{z=t_c}\left(-\frac{\beta_n}{\alpha_m} + \nu_{cyx}\frac{\alpha_m}{\beta_n}\right) + \frac{G_{fxy}\beta_n}{K_x}\lambda_{gn1} + \frac{G_{fxy}\beta_n}{K_y}\lambda_{gk2}\right]C_x C_y + \sum_{m=1,3,..}^{\infty}\sum_{n=1,3,..}^{\infty}C_{mn}\left[\frac{G_{fxy}}{E_{cy}}\phi_y\Big|_{z=t_c}\left(-\frac{\alpha_m}{\beta_n} + \nu_{cxy}\frac{\beta_n}{\alpha_m}\right) + \frac{G_{fxy}\beta_n}{K_x}\lambda_{gn2} + \frac{G_{fxy}\alpha_m}{K_y}\lambda_{gk1}\right]C_x C_y \quad (48)$$

$$q_x = \sum_{m=1,3,..}^{\infty}\sum_{n=1,3,..}^{\infty}\left(A_{mn}\lambda_{gn1} + C_{mn}\lambda_{gn2}\right)C_x S_y \quad (49)$$

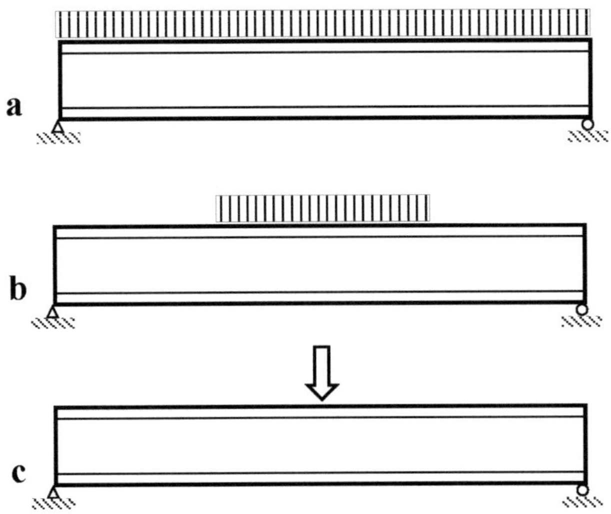

Figure 9.
Three loading cases used for verification, (a) uniform load, (b) central uniform load, (c) central concentrated load.

$$q_y = \sum_{m=1,3,..}^{\infty} \sum_{n=1,3,..}^{\infty} (C_{mn}\lambda_{gk1} + A_{mn}\lambda_{gk2}) S_x C_y \tag{50}$$

$$\begin{aligned}\tau_{cxz} = &\sum_{m=1,3,..}^{\infty} \sum_{n=1,3,..}^{\infty} \int_{z=0}^{z} \phi_x \, dz \, A_{mn} \left[-\alpha_m + \frac{G_{cxy}}{E_{cx}} \beta_n \left(-\frac{\beta_n}{\alpha_m} + \nu_{cxy} \frac{\alpha_m}{\beta_n} \right) \right] C_x S_y \\ &+ \frac{G_{cxy}}{E_{cy}} \sum_{m=1,3,..}^{\infty} \sum_{n=1,3,..}^{\infty} \int_{z=0}^{z} \phi_y \, dz \, C_{mn} \beta_n \left(-\frac{\alpha_m}{\beta_n} + \nu_{cxy} \frac{\beta_n}{\alpha_m} \right) C_x S_y\end{aligned} \tag{51}$$

$$\begin{aligned}\tau_{cyz} = &\sum_{m=1,3,..}^{\infty} \sum_{n=1,3,..}^{\infty} \int_{z=0}^{z} \phi_y \, dz \, C_{mn} \left[-\beta_n + \frac{G_{cxy}}{E_{cy}} \alpha_m \left(-\frac{\alpha_m}{\beta_n} + \nu_{cxy} \frac{\beta_n}{\alpha_m} \right) \right] C_y S_x \\ &+ \frac{G_{cxy}}{E_{cx}} \sum_{m=1,3,..}^{\infty} \sum_{n=1,3,..}^{\infty} \int_{z=0}^{z} \phi_x \, dz \, A_{mn} \alpha_m \left(-\frac{\beta_n}{\alpha_m} + \nu_{cyx} \frac{\alpha_m}{\beta_n} \right) S_x C_y\end{aligned} \tag{52}$$

in which

$$A_{mn} = \frac{\frac{\lambda_{y3}}{\lambda_{y2}} - \frac{\lambda_{y6}}{\lambda_{y5}}}{\frac{\lambda_{y1}}{\lambda_{y2}} - \frac{\lambda_{y4}}{\lambda_{y5}}} \tag{53}$$

$$C_{mn} = \frac{\frac{\lambda_{y3}}{\lambda_{y1}} - \frac{\lambda_{y6}}{\lambda_{y4}}}{\frac{\lambda_{y2}}{\lambda_{y1}} - \frac{\lambda_{y6}}{\lambda_{y4}}} \tag{54}$$

$$\lambda_{y1} = \frac{\lambda'_{z1}}{E_{fx}} - \frac{\nu_{fxy}\lambda'_{z3}}{E_{fy}} - \frac{\phi_x|_{z=t_c}}{E_{cx}} + \frac{\alpha_m \lambda_{gn1}}{K_x}$$

$$\lambda_{y2} = \frac{\lambda'_{z2}}{E_{fx}} - \frac{\nu_{fxy}\lambda'_{z4}}{E_{fy}} + \frac{\nu_{cxy} \, \phi_y\big|_{z=t_c}}{E_{cy}} + \frac{\alpha_m \lambda_{gn2}}{K_x}$$

$$\lambda_{y3} = \frac{2}{a\alpha_m} \frac{2}{b\,\beta_n} \left[\frac{\sigma_{xo} + \sigma_{cxo}\frac{t_c}{t_f}}{E_{fx}} - \nu_{fxy} \frac{\sigma_{yo} + \sigma_{cyo}\frac{t_c}{t_f}}{E_{fy}} + \frac{\sigma_{cxo}}{E_{cx}} + \nu_{cxy} \frac{\sigma_{cyo}}{E_{cy}} \right]$$

$$\lambda_{y5} = \frac{\lambda'_{z4}}{E_{fy}} - \frac{\nu_{fyx}\lambda'_{z2}}{E_{fx}} - \frac{\phi_y\big|_{z=t_c}}{E_{cy}} + \frac{\beta_n \lambda_{gk1}}{K_y}$$

$$\lambda_{y6} = \frac{2}{a\alpha_m} \frac{2}{b\,\beta_n} \left[\frac{\sigma_{yo} + \sigma_{cyo}\frac{t_c}{t_f}}{E_{fy}} - \nu_{fyx} \frac{\sigma_{xo} + \sigma_{cxo}\frac{t_c}{t_f}}{E_{fx}} + \frac{\sigma_{cyo}}{E_{cy}} + \nu_{cyx} \frac{\sigma_{cxo}}{E_{cx}} \right]$$

2.9 Effects of bonding stiffness on the performance of laminates

The complexity of the closed-form solution makes it difficult to ascertain manually the effects of non-rigid bonding on the laminates' responses. Thus, we developed windows-based toolkits to evaluate the obtained complex mathematical formulas [19, 24]. For verification, consider the laminates shown in **Figure 10**. The skins and core properties are: t_f = 0.04 in. (1 e-3 m), E_f = 10^7 psi (6.9 e10 Pa), t_c = 2.0 in. (5 e-2 m), E_c = 2 x 10^4 psi (110 kPa), and G_c = 10^4 psi (69 MPa). Three loading types are considered as shown in **Figure 10**. The laminated panel is under three loading cases.

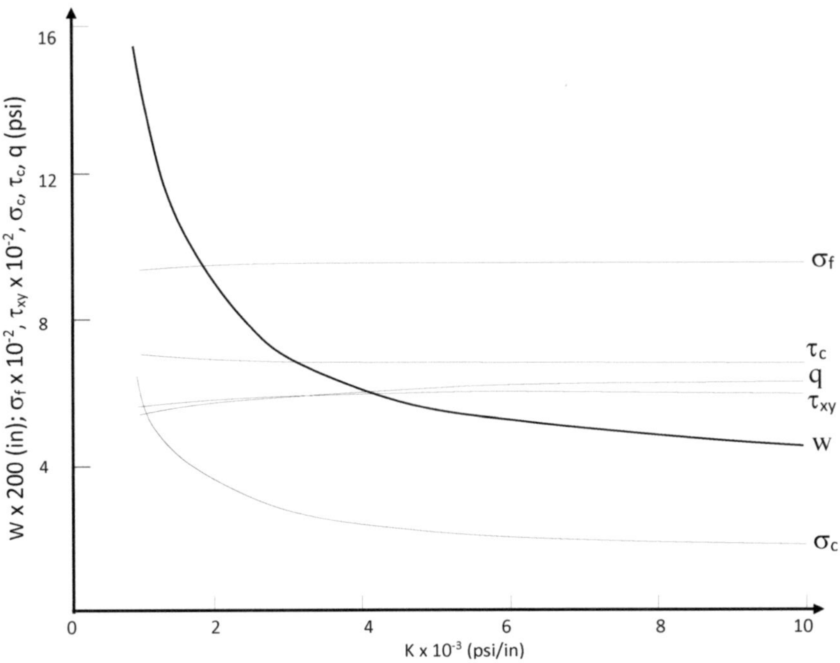

Figure 10.
Bonding effects on deflection and stresses (load case a).

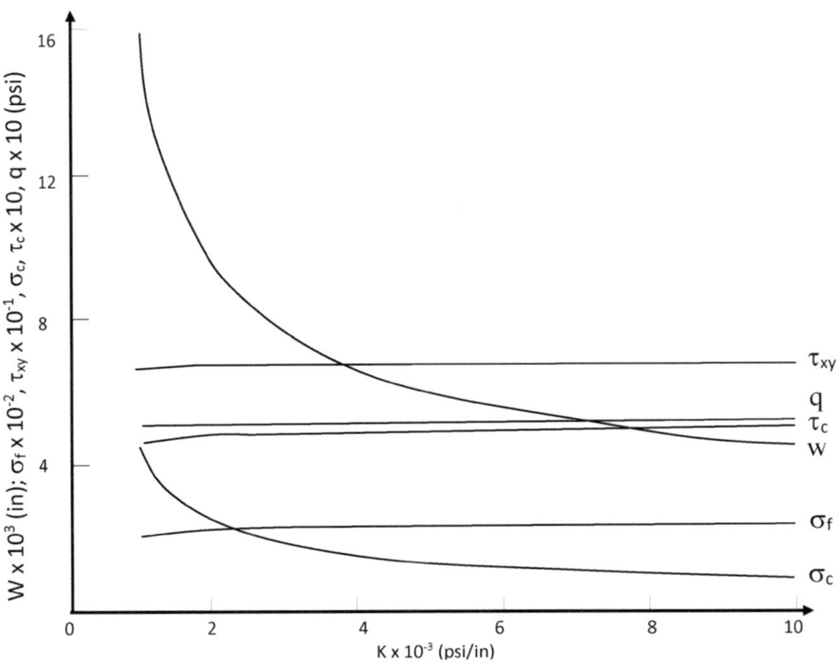

Figure 11.
Bonding effects on deflection and stresses (load case b).

Figures 11–15 show the maximum normal and shear stress, and the deflection obtained for each load. The figures show that the bonding stiffness, K, has significant effect on the deflection in its low range, beyond which the effect is unnoticeable thus the literature common assumption of perfect rigid bonding may

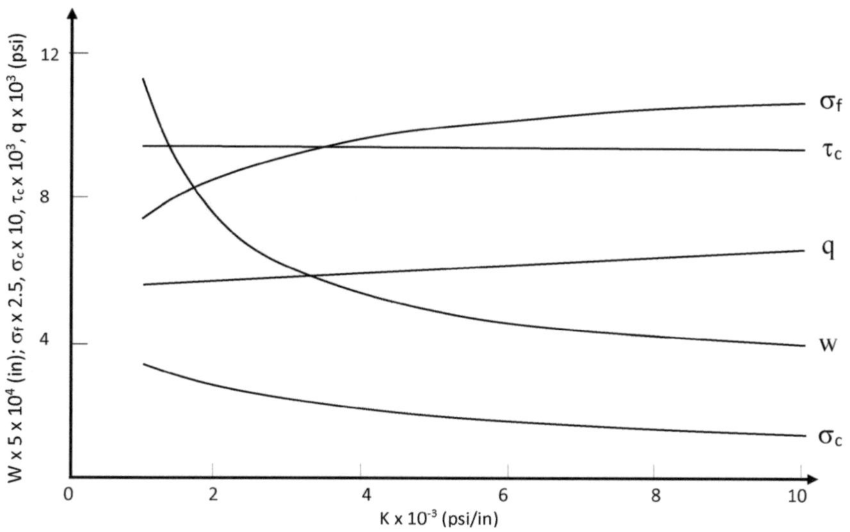

Figure 12.
Bonding effects on deflection and stresses (load case c).

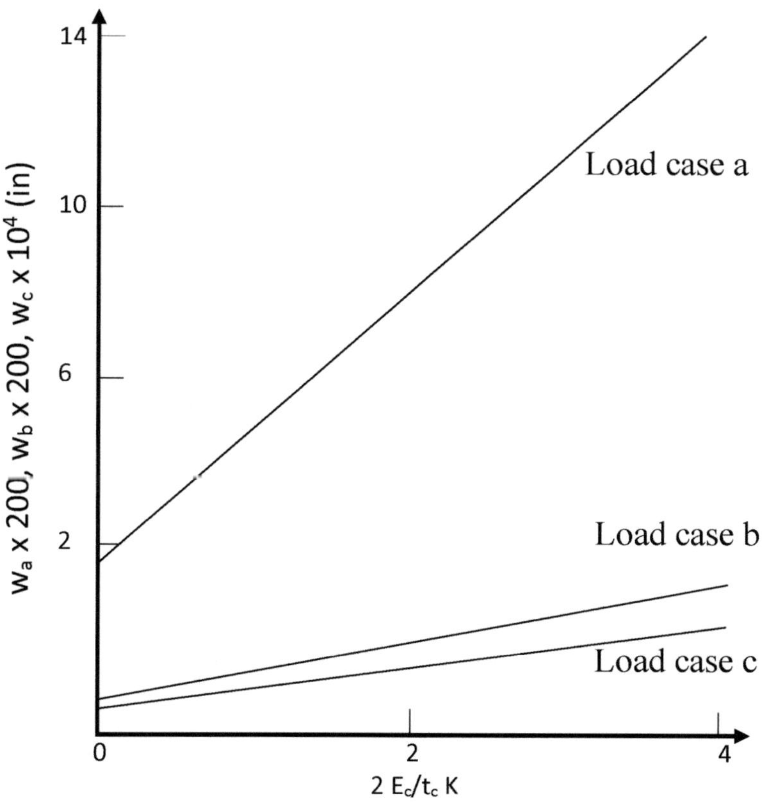

Figure 13.
Effect of (2 E_c/t_c K) on maximum deflection.

apply. To elaborate, the deflection of the panel under uniform load has increased 14 orders of magnitude when the K values changed from 3,000–2,000 psi/in. (8 e8–5.4 e8 Pa/m) to 10–9 ksi/in. (2.7–2.4 GPa/m). Furthermore, the K value showed

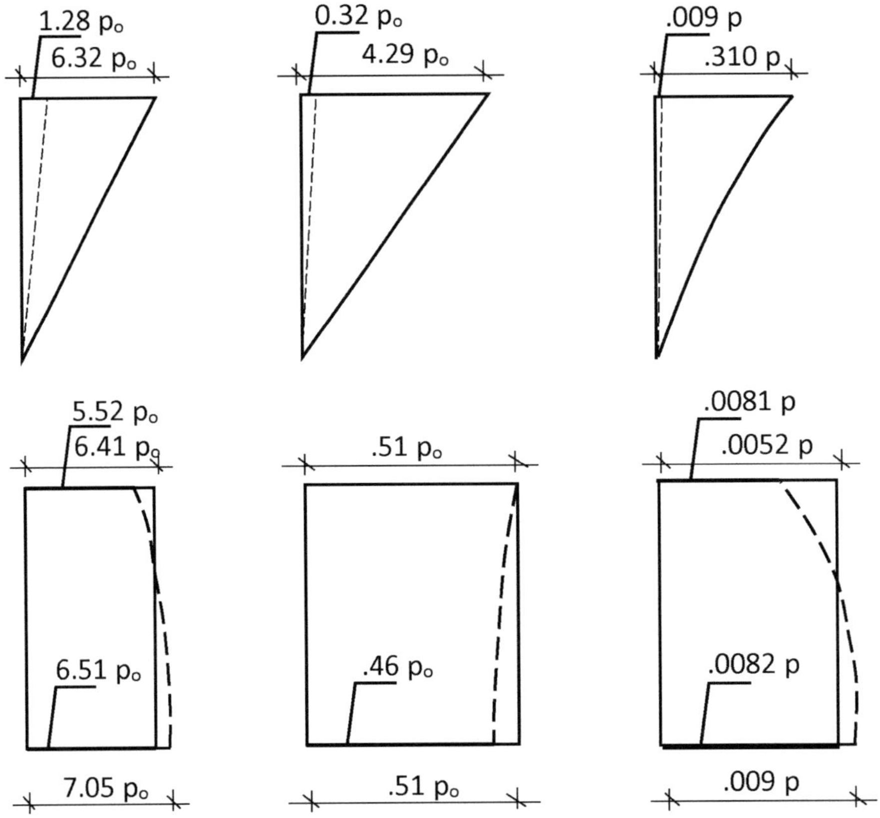

Figure 14.
Bonding effects on stress distributions in Core.

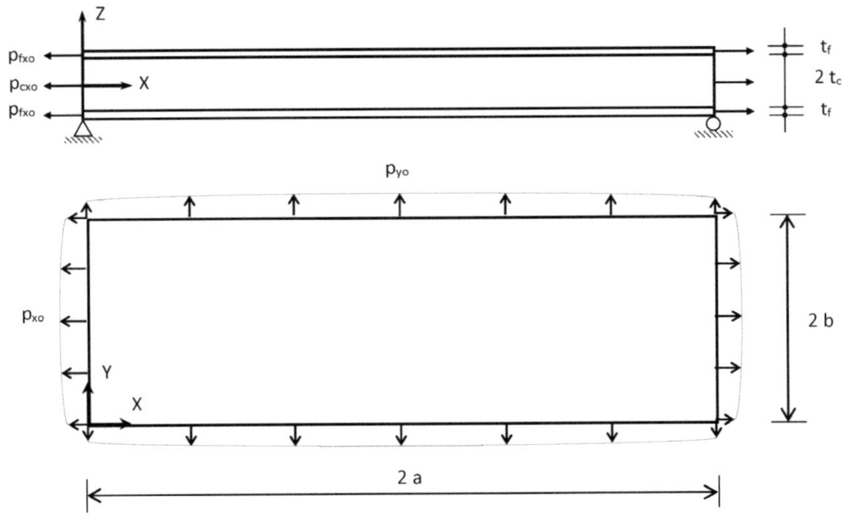

Figure 15.
A laminate under In-plane loads.

insignificant effect on the interlayer shear stress and reduced slightly the core normal stress. The later reduction is compensated by an increase in the skin stress. This discovery is expected because the higher the K value the less deflection, thus less deformation in the thick layer, i.e., the core.

By using the existing rigid-bonding based solutions from the literature [13, 15], normal and shear stresses in the skins, and the transverses shear stress in the core may be determined with an acceptable margin of error. Nonetheless, the bonding stiffness should be included in the analysis whenever the deflection is the quantity of interest. This is an important point. To illustrate this important finding, consider the laminates in **Figure 10**. For a bonding stiffness of 4 ksi/in. (1.0 GPa/m), the deflection and stresses in the laminates are obtained with the present theory as ω = 0.0313 in. (8e-4 m); σ_f = 950 psi (6.6e6 Pa); τ_{xy} = 592 psi (4e6 Pa); and τ_c = 6.7 psi (46 kPa). The corresponding values as obtained using rigid bonding are ω = 0.0168 in. (4e-4 m); σ_f = 939 psi (6.5e6 Pa); τ_{xy} = 610 psi (4.2e6 Pa); and τ_c = = 6.6 psi (4.5 kPa). Comparison of these two sets of results indicates that the deflection determined by the existing theory is underestimated by 46% whereas the stresses of the two are comparable. There is another takeaway from this solution. This study has yet to bring up another important discovery. From an economic perspective, a rigid, i.e., expensive, adhesive would be unnecessary for the manufacturing of the panels if the core is made of soft material such as foam. The converse would be unwise. Because the core and bonding together contribute to the integrity and overall performance of the laminates, the ratio of core-to-bonding stiffness is introduced, and its effect is shown in **Figure 14**. The figure agrees with this discovery.

Laminated panels in building construction undergo various types of loads. The previous section presented a model for panels under flexural loads. In-plane loads are very common in walls and gable roof applications. The analytical formulation for this type of panels is similar to the previously presented one for flexural loads. Thus, it will be summarized in the following section to highlight the differences between the two formulations. The laminate shown in **Figures 9** and **16** is subjected

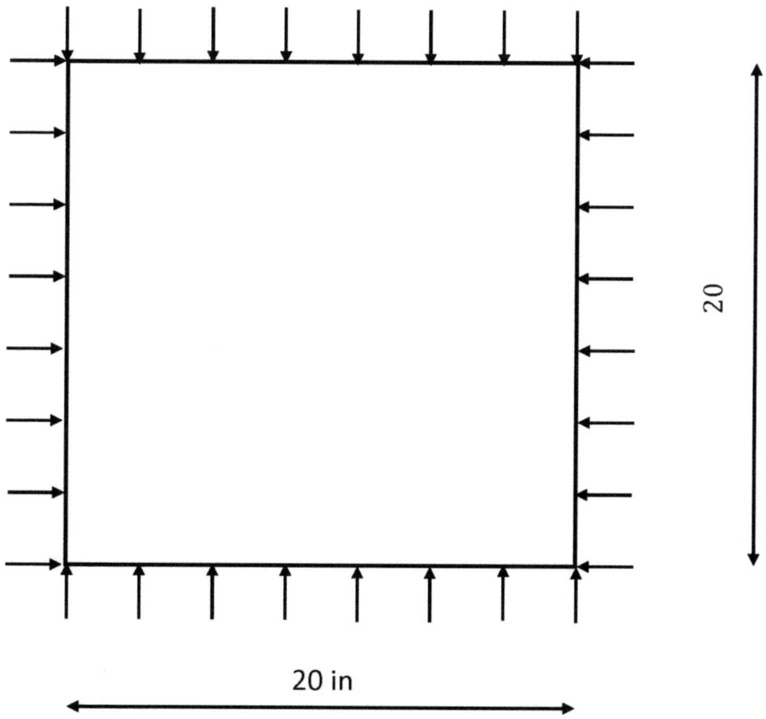

Figure 16.
A plane view of a panel under biaxial edge load.

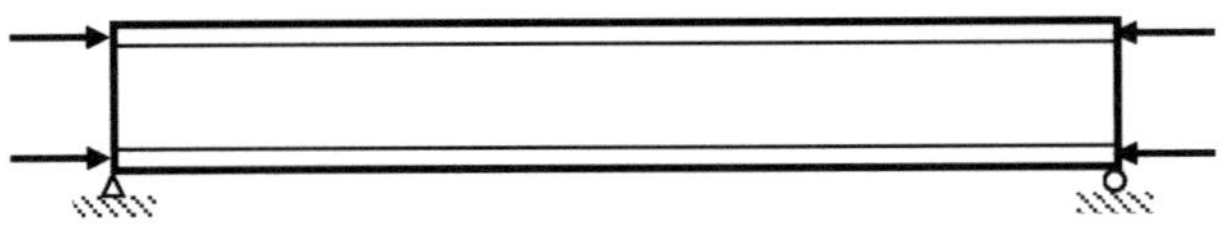

Figure 17.
A biaxial edge load applied to faces.

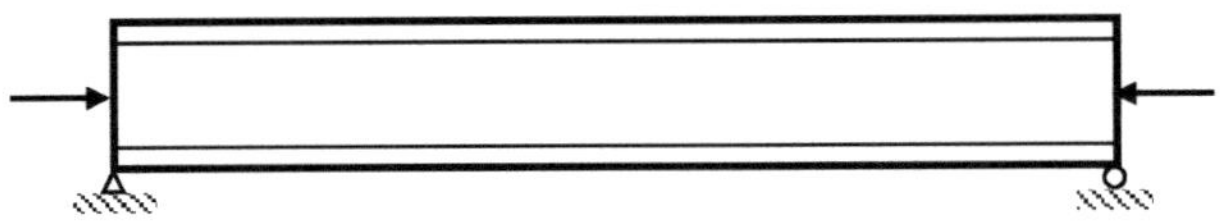

Figure 18.
A biaxial edge load applied to Core.

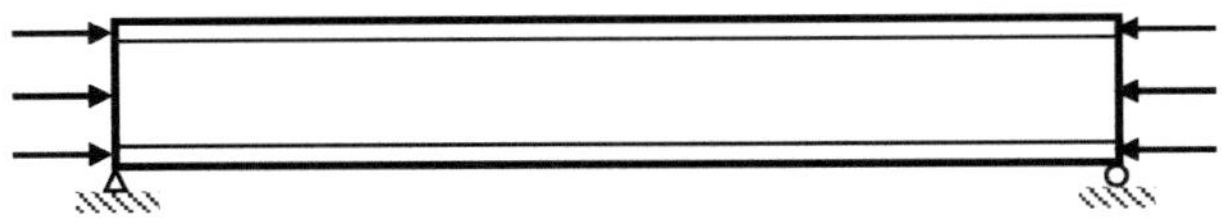

Figure 19.
A biaxial edge load applied to faces and Core.

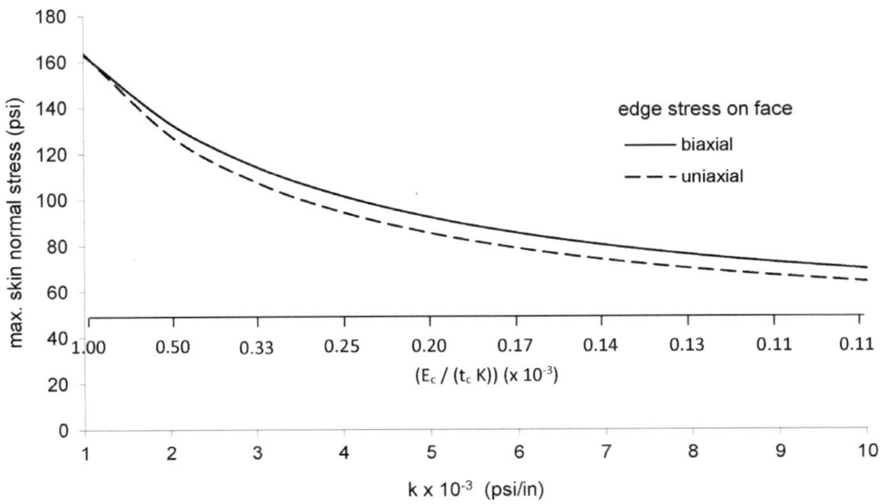

Figure 20.
Effects of bond stiffness on face Normal stress due to face edge load.

to uniform in-plane loadings: a biaxial stress $\sigma_{fxo} = \sigma_{fyo} = 208.3$ psi (1.4 MPa), and a uniaxial stress $\sigma_{fxo} = 208.3$ psi (1.4 MPa). **Figures 17** and **18** show the skin and core loading, whereas **Figure 19** shows the superimposed loading. A wide range of isotropic adhesive stiffness was used, varying from ksi/in (0.25GPa/m) to 10 ksi/in (2.5 GPa/m). This range covers non-rigid to rigid stiffnesses. The normal stress in the faces at the panel center and the shear stress in the faces at the panel corner are shown graphically in **Figures 20–25**.

As under bending loads, the figures, the bonding stiffness affects the response of the laminates. For example, the face normal stress, σ_f, is greatly affected by the K

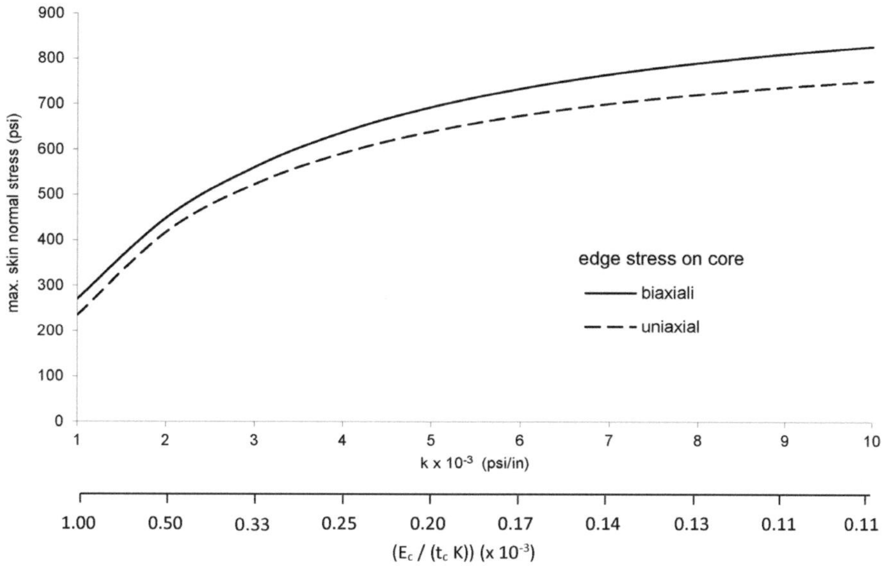

Figure 21.
Effects of bond stiffness on face Normal stress due to Core edge load.

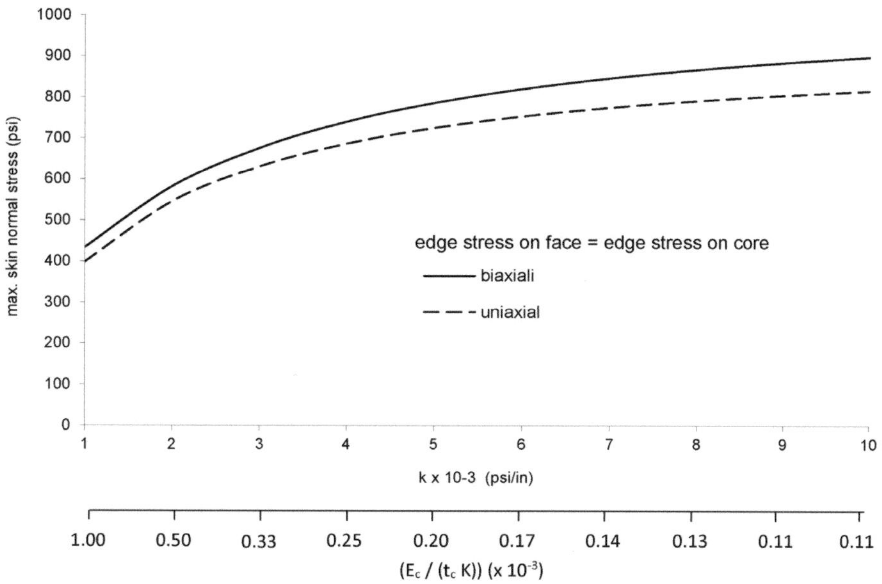

Figure 22.
Effects of bond stiffness on face Normal stress due to face and Core edge load.

value in its lower range; but beyond a certain level of stiffness (which varies from panel to panel), the bonding layer can be practically considered as rigid. A change in K-value for example from 1 to 2 ksi/in (0.25–0.50 GPa/m) induces a stress decrease almost 6 times in the uniaxial case and 5 times in the biaxial case greater than when K changes from 9–10 ksi/in. (2.25–2.50 GPa/m). The changes are 24% and 27% due to uniaxial core and combined edge loads, respectively, 32% and 22% due to biaxial core and combined edge loads, respectively. In all in-load cases, the face and interlayer shear stresses are not affected significantly by the bonding stiffness.

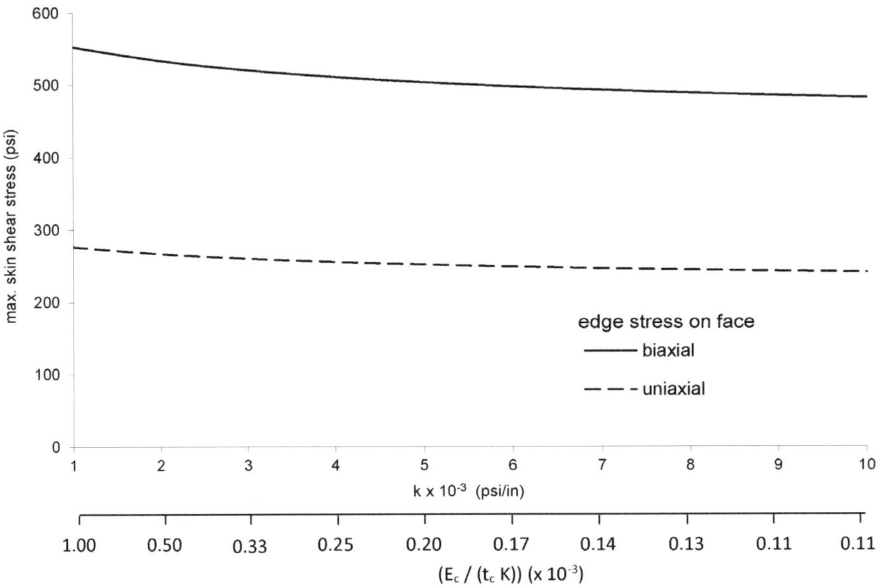

Figure 23.
Effects of bond stiffness on face shear stress due to face edge load.

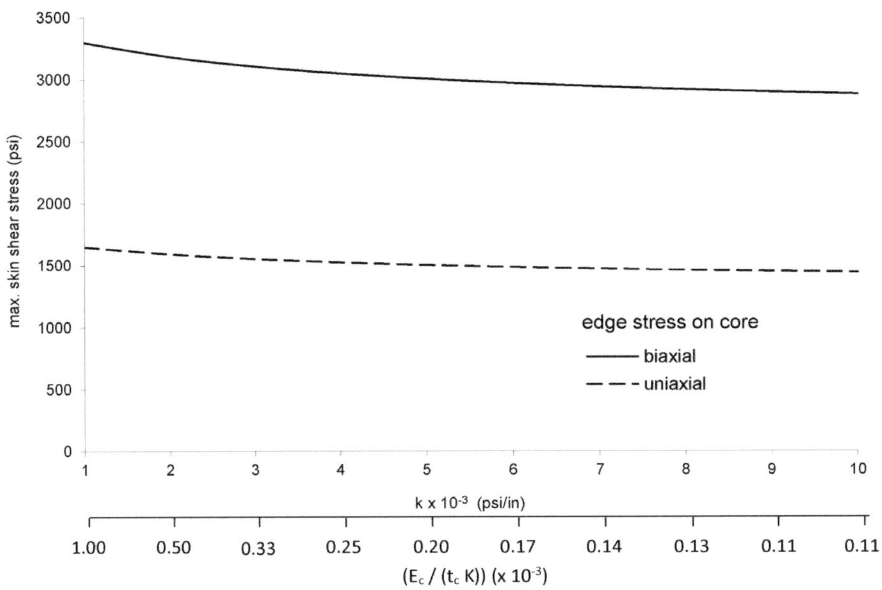

Figure 24.
Effects of bond stiffness on face shear stress due to Core edge load.

The absence of transverse loads yields no transverse shear which coexists with in-plane shear.

As found in panels under flexural loads, a panel made of adhesive with high stiffness and soft cores is uneconomic, and the converse is unsafe. Thus, the applications of the theories that assumed perfectly rigid bonding should produce acceptable performance results if the bonding stiffness is high and the core material is compatible with the adhesive.

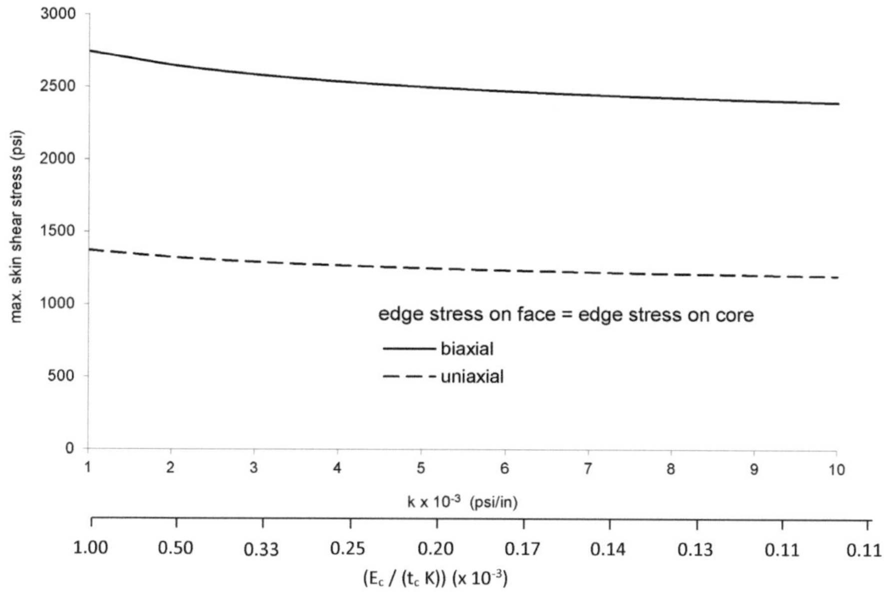

Figure 25.
Effects of bond stiffness on face shear stress due to Core edge load.

3. Engineered gussett plates in IN timber systems

3.1 Introduction

Metal gusset plates are used in timber light frame construction since 1955 to create structural joints. The concept of perforated gussets joins the timber members without the use of nails, screws, or glue. The building construction industry is familiar with the timber framing that use that concept, thus is empowered to meet myriad of challenges including complex blueprints and structural joints [18]. However, the incorporated laterally unsupported areas in the gussets are subject to out-of-plane deformation due to excessive forces. Currently, the literature incorporates some experimental data for the metal connectors [1, 13, 18], and empirical design recommendations [13, 20]. To our knowledge, the out-of-plane deformation of the gussets was not engineered so far. This chapter introduces for the first-time relevant design aids to fill the relevant voids in the existing literature [1, 13, 16, 20]. To provide a well-balanced presentation, we presented the analytical background of the problem at hands, upon which a design procedure was established and introduced.

3.2 Framework analysis

Any structural analysis requires material and geometrical properties of the members that constitute the framing system. The constitutive relations of timber are well established and applied since the appearance of Hooke's law in 1676. In this study, we considered timber as a 3D orthotropic material for which the required mechanical properties are adopted from Ref. 25. The gusset-timber in-plane and out of plane interactions are complex to simulate [1, 13, 20]. This study idealized the in-plane and out of plane interactions using spring elements. For this purpose, the stiffness matrix is

$$[K] = \begin{bmatrix} K_{11} & K_{12} \\ symm & K_{11} \end{bmatrix}$$

where K_{ii} is a 6x6 diagonal sub-matrix with the axial and shear constants, whereas K_{ij} is zero matrix.

Because of the inherited complexity of the problem at hands, the finite element method is deemed a direct and simple method to harness all the previously described component developed for our approach. **Figures 26–28** show FE models that represent various timber joints. The material properties for timber and gusset are: E_{gusset} = 30 ksi, $E_{l\text{-}timber}$ = 1.6 ksi, $E_{t\text{-}timber}$ = 0.08 ksi, $E_{r\text{-}timber}$ = 0.108 ksi,

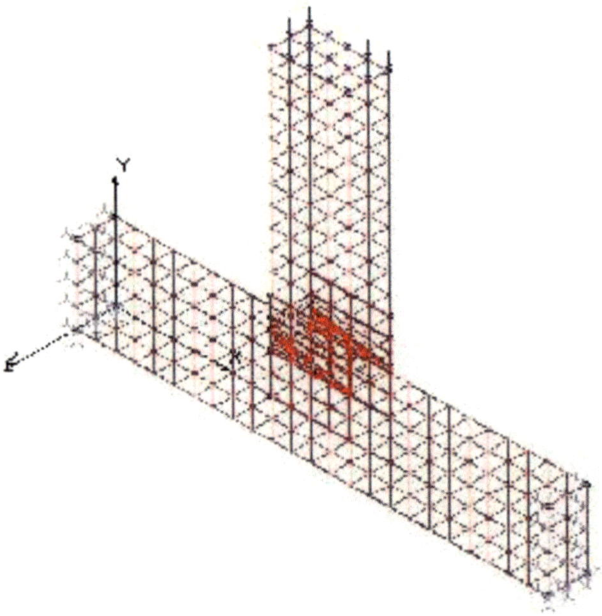

Figure 26.
A screen view of the framework analysis - a.

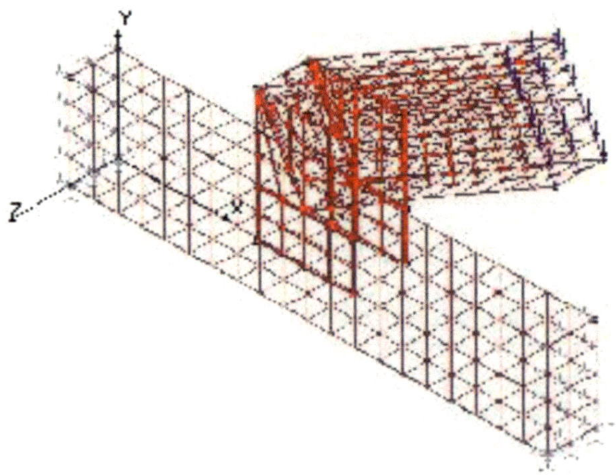

Figure 27.
A screen view of the framework analysis - b.

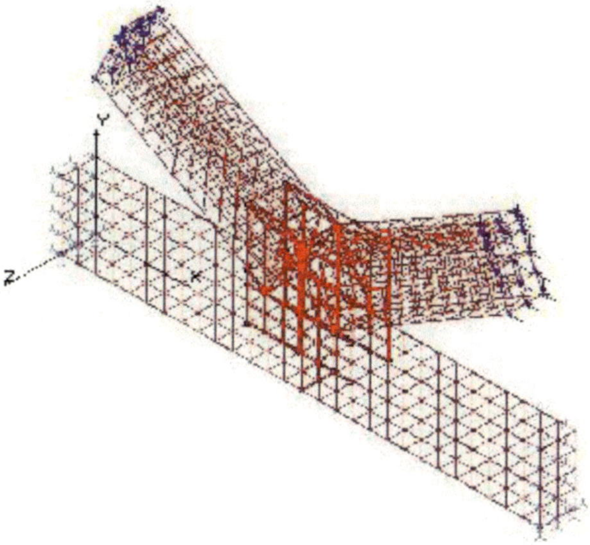

Figure 28.
A screen view of the framework analysis - c.

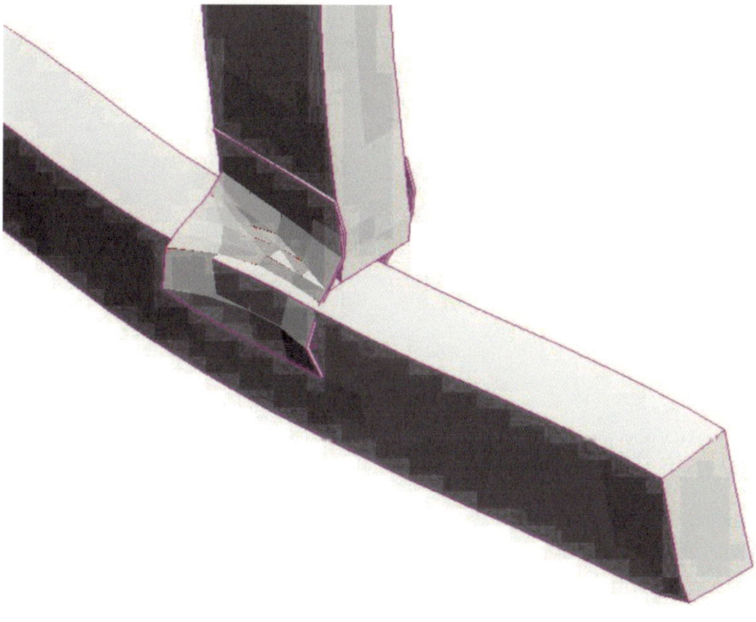

Figure 29.
*Out of plane deformation of the gusset in **Figure 26**.*

G_{lr} = 0.102 ksi, G_{lt} = 0.124 ksi, G_{rt} = 0.111 ksi, $?\nu_{lr}$ = 0.29, $?\nu_{lt}$ = 0.45, $?\nu_{rt}$ = 0.39, $?\nu_{tr}$ = 0.37, $?\nu_{rl}$ = 0.04, $?\nu_{tl}$ = 0.03, where E is modulus of elasticity, n is Poisson's ration; l, t and r stand for longitudinal, tangential and radial directions, respectively. **Figures 29–31** show the corresponding deformed shapes. As shown in the figures, the gusset plates undergo bending and withdrawal deformations.

Figure 32 shows the performance of the gusset in **Figure 26**. The figure shows 80% and 70% drops in the critical load (the load at which out of plane deformations occur) under compression and shear, respectively. Under chord and gap shears, the critical load decreases and increases by 70% and 27%, respectively. This finding

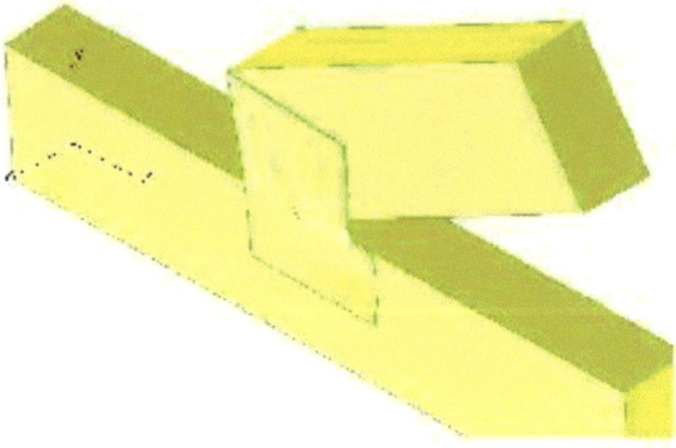

Figure 30.
*Out of plane deformation of the gusset in **Figure 27**.*

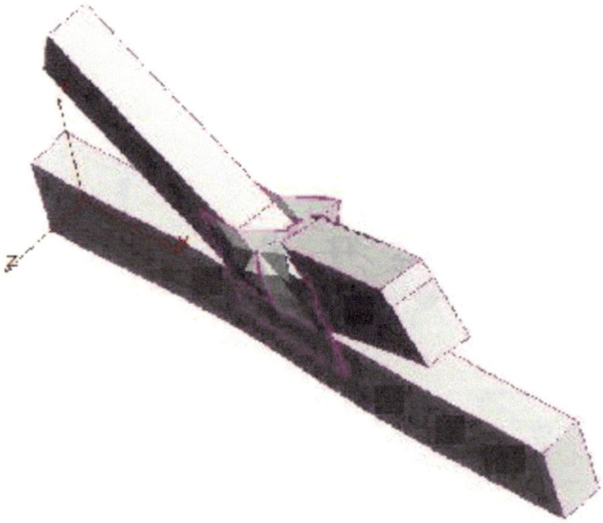

Figure 31.
*Out of plane deformation of the gusset in **Figure 28**.*

suggests that the plate is weaker under than compression for areas smaller than 2.0 in$^{2.}$ and the converse is true for larger areas. For practical applications, the smaller ratio of gap/shear length the higher the critical load. For example, the load increases by 50% when the ratio is ¾ rather than 4/3. For all applications, the smaller the gusset unsupported area or its thickness the higher the critical values, and this agrees with conventional theories [17]. One of the takeaways from this analysis is that chord and web shears rather than compression [13] are the key drivers for the gusset instability. In addition, an answer the question of what constitutes effective gusset area can now be answered. An area of 1.0 in^2 would be conservative regardless of thicknesses whereas caution is advised when using an area larger than 2.0 in^2. However, designers can use large areas in combination with thicker plates. In general, the design needs not to depend on one variable only, gap, or chord length, or gusset area. This is another takeaway. To validate our analytic approach, we used the data obtained by the truss plate institute from testing various connections made of Southern Pine. **Figure 33** compares both results. The figure shows close agreement between the critical values.

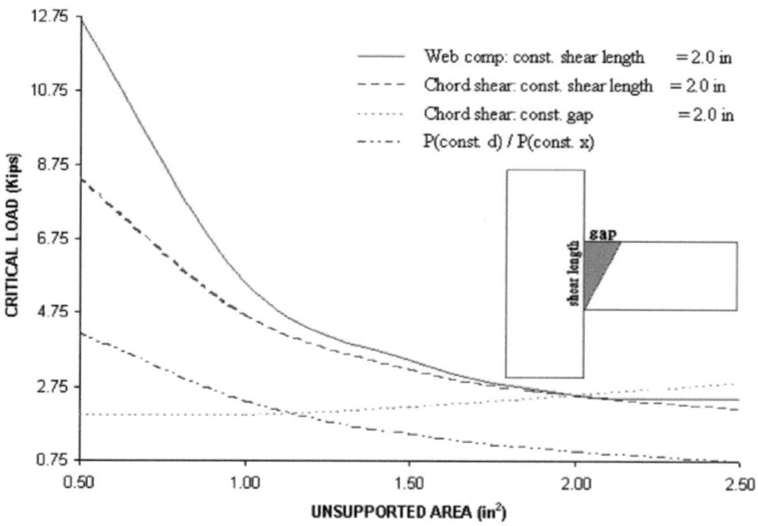

Figure 32.
Effects of gap and shear length the plate in ***Figure 26***.

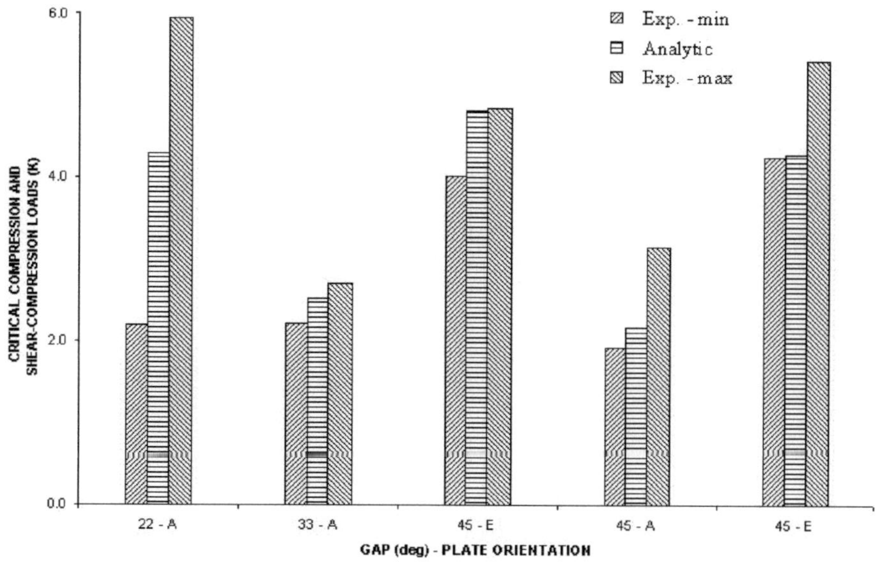

Figure 33.
Comparison of experimental critical loads for the gusset in ***Figure 26***.

Based on the confidence established in the analytic formulation, we then moved on to gusset plates with three laterally unsupported areas. **Figures 34–36** show the deformations obtained for areas A, B, and C. It should be noted that these figures show the gusset plate only. In addition, the web force compresses area A and shears areas B and C.

Figure 37 shows that area B deforms before A under chord shears. This occurs only under web compressions for areas up to about 1.5 in^2, afterwards this sequence reverses. When area B is between 0.5 in^2 and 1.0 in^2, the critical load decreases by 54% and 36% under compressions and shears, respectively. When the area is between 1.0 in^2 and 2.0 in^2, the decrease is 28% under compression and 26% under shear. These findings are not accustomed in practice where only web compressions

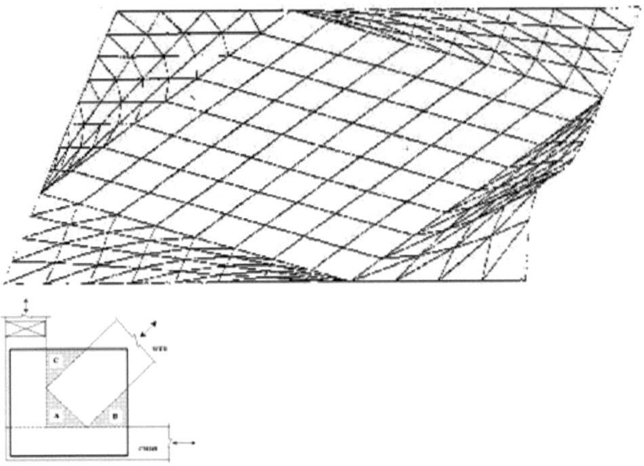

Figure 34.
*Deformation of area a in the joint in **Figure 34**.*

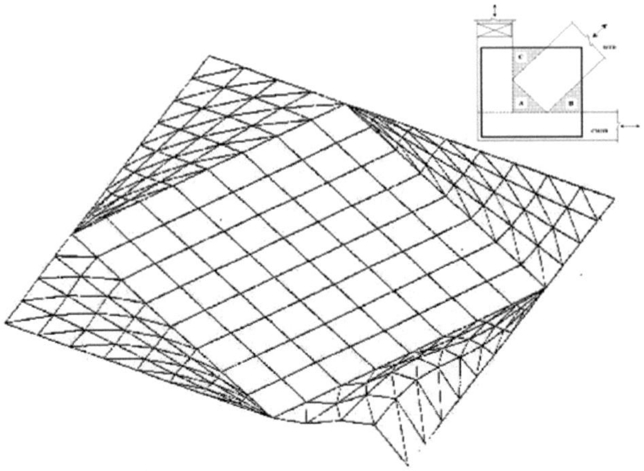

Figure 35.
*Deformation of area B in the joint in **Figure 34**.*

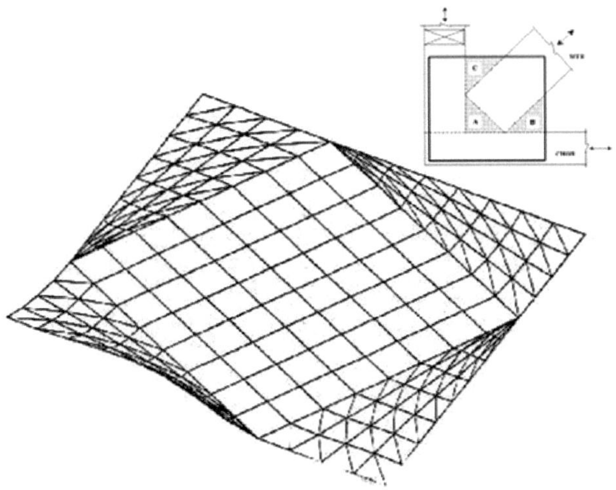

Figure 36.
*Deformation of area C in the joint in **Figure 34**.*

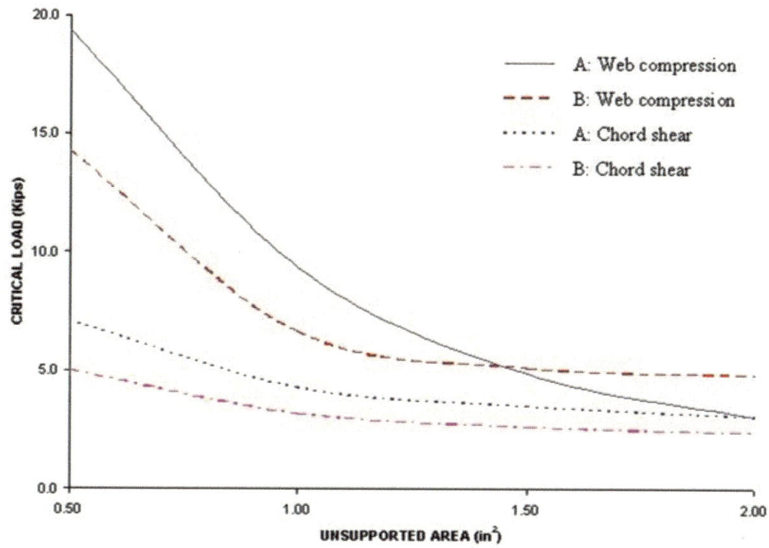

Figure 37.
Effects of unsupported areas on shears and web compressions (0.030 in).

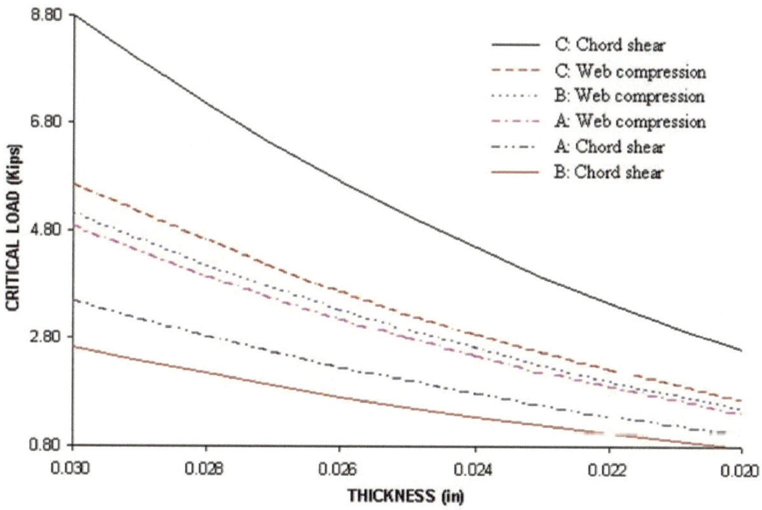

Figure 38.
Comparison between web compression and chord shear critical loads of areas a, B and C (1.5 in²).

is the driving factor for deformation. **Figures 37** and **38** affirms that the chord shear has more effects on the deformation of the gusset than compression. In agreement with the analysis conducted for single area and fundamentals of engineering mechanics, the smaller the gusset or the larger its thickness, the larger the critical values.

3.3 Design aids for gusset plates

To bring the outcomes of our investigation to everyday practice, **Figures 39–42** show design curves. To use any of the figures, area A or B is first located on the horizontal axis, then a vertical line is drawn to intersect with the curve with the

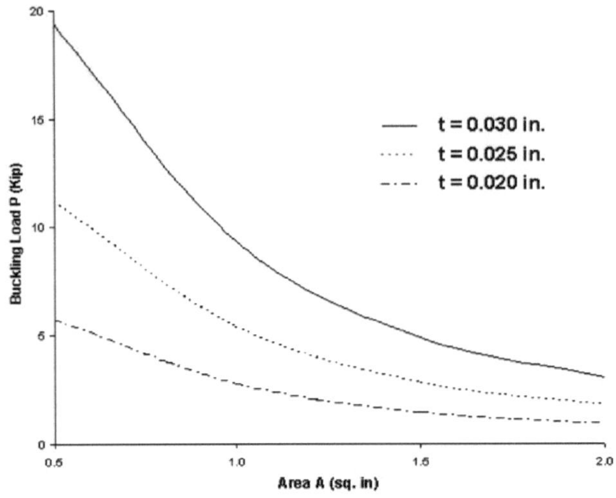

Figure 39.
Design critical loads for gusset under web compression.

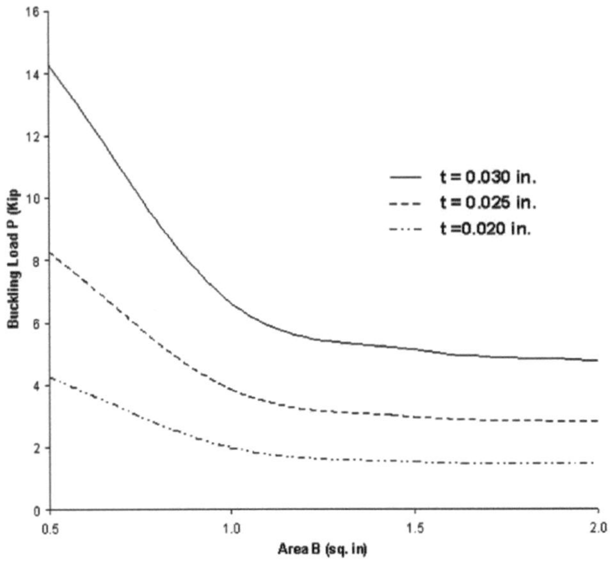

Figure 40.
Design critical loads for gusset under web compression.

design thickness. A horizontal line is finally drawn to intersect with the vertical axis at the critical load.

4. Applications of NDT for timber construction

4.1 Introduction

Shelters represent one of the basic human needs and residential buildings meet this need. References [1, 13, 16, 20] depict the state of deteriorated buildings and the methods used for assessing the status of their structural condition. One could

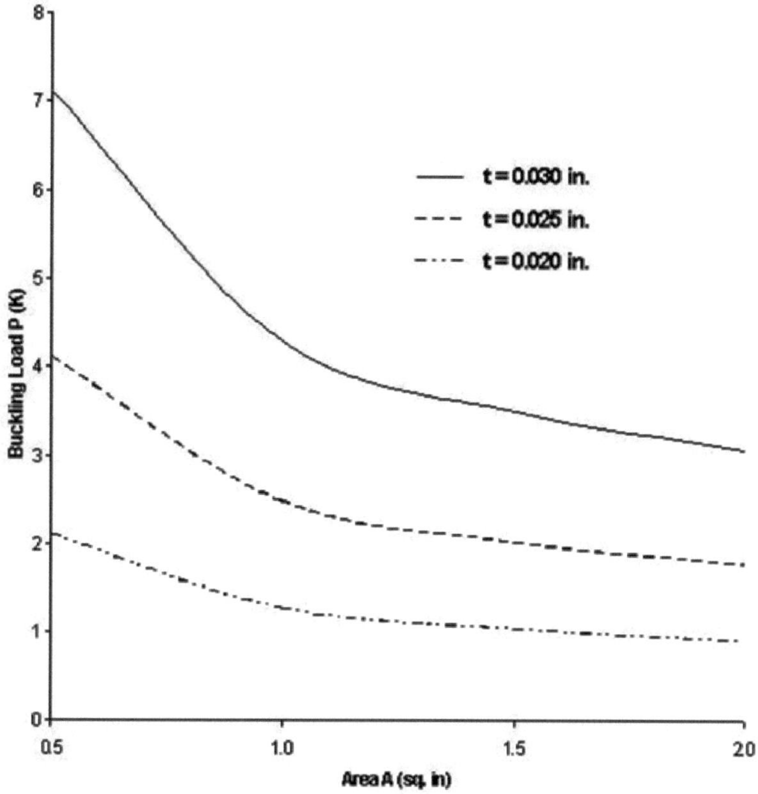

Figure 41.
Design critical loads for gusset joints under chord shear.

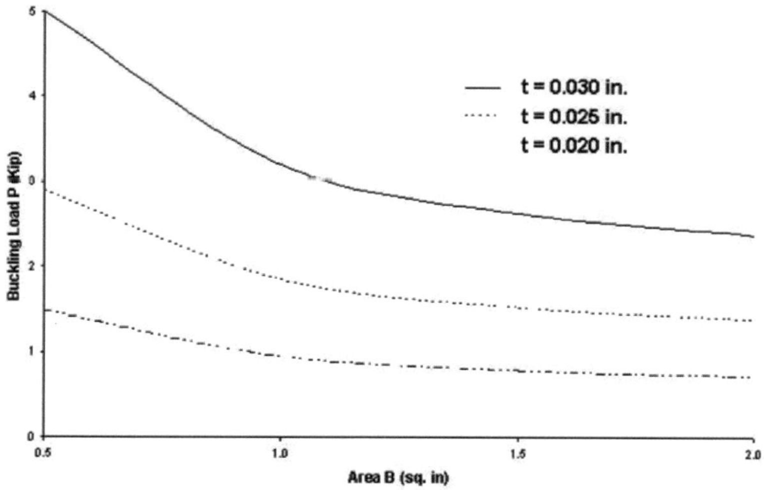

Figure 42.
Design critical loads for gusset joints under chord shear.

see in these sources and many others the complexity of the technical tasks involved in the repair projects.

In general, the need to determine the true structural status of load bearing members is motivated by a number of circumstances such as:

- The existence of defects in the system.
- Damage after natural events such as earthquakes.
- The change in the function of buildings.
- New regulations.

The assessment of the actual structural condition of the load bearing members is a necessary step in all engineering projects. This step is not easy to take because of the architectural configurations and the framing systems. Thus, the assessment of the actual condition of a component comprises in-situ tests and procedures as described by the authority in charge of the projects. In general, tests are expensive and not simple to perform, let alone feasible to conduct. In addition, monitoring procedures are expensive and cumbersome. Thus, non-destructive approaches lend practical ways to overcome these obstacles. From this perspective, a new non-destructive approach has been adopted in this chapter for fast yet reliable outcomes.

4.2 Description of the problem

The engineering analysis and design of construction systems start with analytical models to be solved for the quantities of interest. The models require the geometrical configuration, the applied loads, pertinent mathematical theories or methods, experimental data if needed, and algorithms to execute the procedure. The outcomes are then used for remedial solutions of fragile systems, retrofitting, rehabilitation, etc.

The modeling of old buildings is, in general, more difficult and less reliable than in the case of new buildings. This is due to several factors, such as:

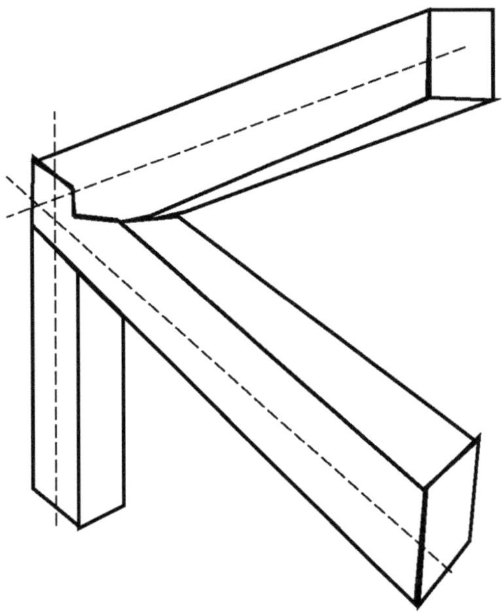

Figure 43.
A typical structural joint.

- The geometrical configuration.
- The uncertainties of the materials properties.
- The invisible impacts of past events such as earthquakes.

Additional factors that compound the analytic modeling include the need for great experience about the structural behavior of building systems, local damages such as cracks, the actual constituent relationships; the alterations over time such as the creation of openings or the removal of components (headers, walls, etc.) or the increase in the height of some spaces, etc. In general, the structural and mechanics procedures are not simple and in most if not all applications require experience. Consider for example, the joint shown in **Figure 43**. In any analytical procedure, it is commonly defined as pin or rigid. Any definition is a mere judgment because true connections are neither of these theoretical cases, and in fact fall in between. The question is: how could a practitioner define the real condition? The answer is essential because the outcomes of the modeling depend on this definition. The rigorous procedure introduced in this chapter contributes to a reliable answer.

4.3 Formulation

To simulate realistically a structural joint, we use semi-rigid connections. Such a joint is neither pin with zero stiffness nor rigid with infinite stiffness. **Figure 44** shows a beam element with end stiffnesses K_i and K_j. The member end forces-displacement and mass relations are expressed in the literature in the form of a 6-by-6 matrices in which the following parameter is used

$$C_i = \frac{L\,K_i}{E\,I + L\,K_i} \tag{55}$$

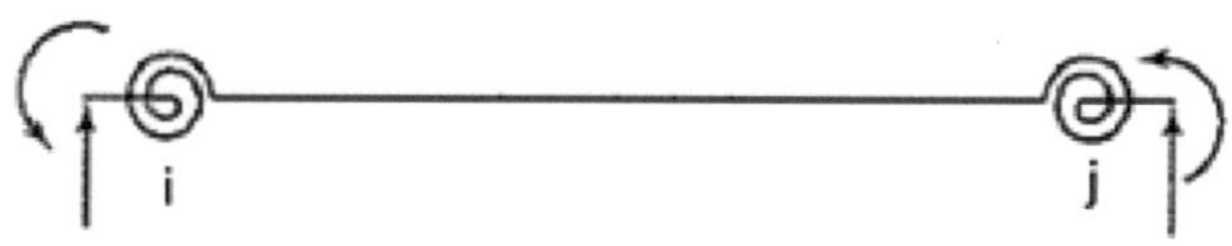

Figure 44.
Frame element with semi-rigid connections.

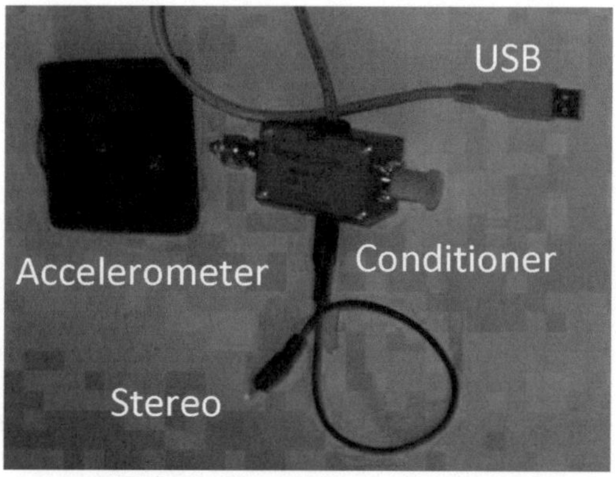

Figure 45.
Signal conditioner.

in which L is the length, I is the sectional moment of inertia, and E is the modulus of elasticity of the material. The structural dynamic governing equations of this are

$$[M]\{\ddot{x}\} + [K]\{x\} = 0 \tag{56}$$

$$\left|[K] - \omega^2[M]\right| = 0 \tag{57}$$

where x is the joint rotational displacement, and ω is the fundamental frequency which can be measured using a setup as shown in **Figure 45** in conjunction with FFT analysis.

Once the geometrical and material properties of the member, and the FFT outcome are obtained, Eq. 57 can be iteratively solved for K_i and K_j. The solution process is referred to in the literature as reverse engineering. It is an iterative process in which initial connections are provided then iteratively updated until Eq. 57 is satisfied.

Putting the described direct, easy to perform, and inexpensive procedure together, this chapter suggests the following new complete protocol for the structural assessment of the joint stiffness of an actual frame member:

1. Measure the geometrical properties of the member.

2. Define the elastic properties of the member' material.

3. Use a set up as shown in **Figure 45** to detect dynamic signals on which use FFT.

4. Solve Eq. 57 for K_i and K_j.

To verify the method introduced in this chapter for the assessment of joint stiffnesses, selected data from the literature is used. NASA [1, 17] applied the finite element method, which is an advance method and requires experience, to investigate many joints from pin to fixed conditions, and the results are shown in

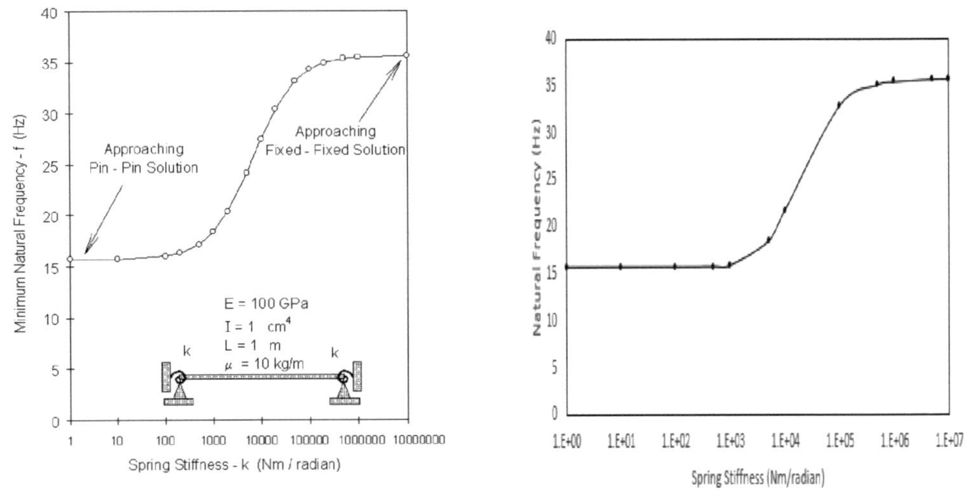

Figure 46.
Comparison between the frequency-rotational stiffness relation determined by NASA and this chapter. (a) NASA, (b) This chapter.

Figure 46a. By using NASA's data, we solved the same problems according to our method, and the results are shown in **Figure 46b**. **Figure 46a** and **b** show no significant difference. There was no need for experience and fast results were obtained. This agreement has built confidence in the accuracy and validity of the introduce method and procedure for buildings construction.

5. Knowledge-based cross laminated timber IN building construction

5.1 Introduction

In general, cellulose materials such as timber have been used in construction since mankind started to build shelters. In the broadest sense, timber is characterized by low weight, prefabrication ability, astatic, acoustic properties, and potential recyclability. The mass timber (MTC) refers to framing system in which timber panels are used for the load bearing components. The products in this category includes cross-laminated (CLT) timber, the nailed laminated timber (NLT), the glued-laminated timber, the doweled-laminated timber (DLT), the structural composite lumber (SCL), and the timber-concrete composites [13]. A 2017 study [20] has doubted the MTC's environmental advantages and even if they are known in induvial construction projects.

According to the CLT Handbook [1], if we perhaps form a hypothesis about likely comparative performance of CLT and assume that CLT has a smaller manufacturing carbon footprint than concrete and steel, it would follow that CLT would perform similarly given that it would have more timber available for landfill storage at end of life. Another study conducted by the Pacific Northwest Building Resilience Coalition [2] in 2018, suggests that the cost of the structural frame for the CLT option is greater than that of the cast-in-place reinforced concrete option, and the concrete in an evaluated ten-story building is less susceptible to floor under dynamic loads, and to excessive lateral drift under wind and seismic loads. The author of this chapter advises caution for judgments when using the available literature. Nonetheless, there are advantages in using timber products but some of the reported disadvantages from MTC include [16] softer materials due to cross laminations, rolling shear, limited dimensions, dimensional stability, fabrication, heavy weights, [17] cost compared to steel and concrete, restrictions on building heights, cost of utilities, cost of design, and cost of transportation. However, as more production and demand will hopefully drive down these costs and increase competitiveness. Some of the factors that affect the environmental benefits of MTC include [20] carbon sequestration rate, timber type, manufacturing processes, material size, equipment and fuel type, regulations, potential of water and insect damages, deconstruction, to name a few examples. The issue of environmental benefits assessment is beyond the scope of this chapter.

Cross laminated timber is an engineered timber category of the MTC and is fabricated by adhering and compressing timber layers in perpendicular grain orientations to form a solid panel. Timber layers are glued together on their wide faces. CLT technology was invented and developed in central Europe in the early 1990's and since then it has been gaining increased popularity in residential and non-residential applications. Recently, CLT was introduced also overseas, in North America, Australia and in New Zealand. A number of production plants have been established or they are proposed to be built in these countries.

The manufacturing process of CLT includes primary lumber selection, lumber grouping and planning, layers cutting to length, adhesive application, panel lay-up, assembly, product curing, surface machining, and marking and packaging. Each of these steps includes several sub-steps and pertinent details can be found in the literature. The manufacturing requirements are well established and practiced, and available in many international sources in the literature.

The European quality control system requires following specific guidelines for production and testing before obtaining technical approvals. In USA, the American National Standards Institute approved the Standard for Performance-Rated Cross-Laminated Timber. These standard covers manufacturing, qualification and quality assurance. The list of other governing organizations includes the American Wood Council, Canadian Standards Association, International Code Council International Building Code, and the Engineered Wood Association.

Internationally, the list of pertinent organizations includes the International Organization for Standardization, American Society for Testing and Materials, New Zealand Standard Timber Structures Standard, and the European Standards.

5.2 Structural strengths and serviceability

The CLT panels act similarly to sandwich construction. The layers in the longitudinal direction carry the bending stresses whereas the transverse layers provide thickness for high inertia and integrate the layers. In this configuration, the transverse shear is carried by all layers. CLT panels have reinforcing effect for prevention from brittle failure modes such as splitting and increases strength capacity of connections. The cross-laminating process provides improved dimensional stability to the product which allows for prefabrication of long and wide panels. There are different methods to connect the panels to each other such as single or double wooden splines, dowels, and mechanical fastener as nails, screws, bolts, split rings, and shear plates.

At this time, this seed R&D focuses on the development of a KB new toolkits for CLT only using Matlab, Python, and VBA for the CLT structural analysis, calculation, design according to the US standards. The algorithms presented here can be used as a foundation for numerical modeling of CLT structures. This step enables the user to model a CLT structure, starting from the creation of the geometry and the assignation of numeric entities (beam, shell, etc.) to geometric ones, having defined the material, and assigning loads and boundary conditions. This is a seed that will blossom to include other structural components not considered at this time.

The governing formulas for the bending and compression design of CLT are [1, 18, 21, 22, 25, 26].

For bending capacity

$$M_b \le F_b S_{eff} C_D C_M\ C_t C_L \tag{58}$$

For shear capacity

$$F_{s\ or\ v}(Ib/Q)'_{eff} = C_M C_t F_s (Ib/Q)_{eff} \tag{59}$$

For compression capacity

$$P_c \le F_c A_{parallel} C_D C_M\ C_t C_P \tag{60}$$

For deflection under uniform load w

$$\Delta_{max} = \frac{5}{384}\frac{w\,L^4}{EI_{eff}} + \frac{1}{8}\frac{w\,L^2}{\frac{5}{6}GA_{eff}} \tag{61}$$

in which

$$S_{eff} = \frac{0.85}{12}\frac{(EI)_{eff}}{E_{major}}\frac{2}{t_p}$$

$(EI)_{eff} = \sum_{i=1}^{n} E_i b_o \frac{t_i^3}{12} + \sum_{i=1}^{n} E_i b_o t_i z_i^2$ = Effective flatwise bending stiffness, lbf-in^2/ft. (N-mm^2/m)

$$(Ib/Q)_{eff} = \frac{(EI)_{eff}}{\sum_{i=1}^{n/2} E_i h_i z_i}$$

$$GA_{eff} = \frac{\left(t_p - \frac{t_1}{2} - \frac{t_n}{2}\right)^2}{\left[\left(\frac{t_1}{2\,G_1 b_o}\right) + \left(\sum_{i=2}^{n-1} \frac{t_1}{G_1 b_o}\right) + \left(\frac{t_n}{2\,G_n b_o}\right)\right]}$$

= Effective flatwise shear rigidity, in lbf/ft. (N/m) in the major strength direction.

M_b = Applied bending moment.
P_c = Applied axial load.
F_b = Reference bending design stress of outer lamination.
S_{eff} = Effective section modulus.
E_i = modulus of elasticity of the i-th layer, in psi (MPa).
$(EI)_{eff}$ = Effective bending stiffness.
GA_{eff} = Effective shear stiffness.
C_D = Load duration factor.
C_M = Wet service factor.
C_t = Temperature factor.
C_L = Beam stability factor.
C_P = Column stability factor per the National Design Specifications (NDS).
t_i = Thickness of laminations in the i-th layer, in (mm).
b_o = CLT width in the CLT major strength direction, in/ft. (mm/m).
z_i = Distance between the centroid of the i-th layer and the neutral axis, in (mm). for the middle layer, zi is to the centroid to the centroid of the top half of that layer.
n = Number of layers.
h_i = Thickness if i-th layer, in (mm).
t_p = Gross thickness of CLT panel, in. (mm.).
$A_{parallel}$= Area of parallel layers parallel to the load.
K_s = Constant for the effect of shear deformation.

The above governing formulas are recognized and applied in practical applications and used, though the author of this chapter has some reservations, for the development of the unparalleled TLC KB toolkit. The dashboard of the TLC Toolkit is shown in **Figure 47**. Like all other novel toolkits introduced in this chapter, it is user friendly and needs the data highlighted in light green color. Once the user clicks the calculate button, the previous formulas are used to calculate the output shown in the white areas. The outputs are needed to calculate the bending and axial capacities, and serviceability parameter of the CLT. These quantities can then be checked using the limits set by adopted design codes.

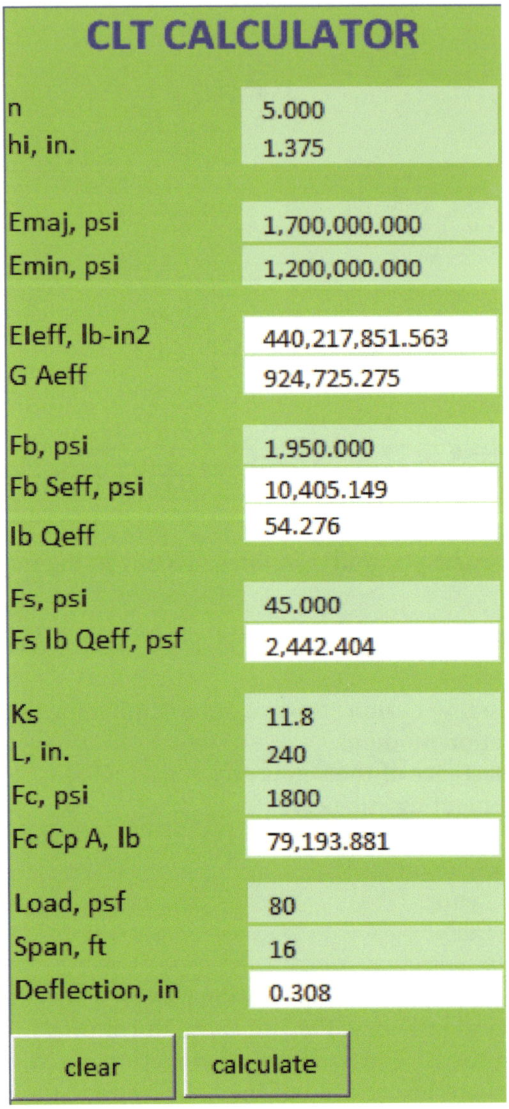

Figure 47.
Dashboard of the CLT toolkit.

6. Knowledge-based toolkits for infrastructure health assessment and monitoring: utility poles/geofoam

6.1 Introduction

Disasters can be prevented or contained by conscious human actions designed. It was this recognition that prompted the author's conceptual shift from the traditional emphasis on response or resistance (post-actions) to disaster monitoring and prediction (pro-actions) as reflected in this section of the chapter. Our carried out works so far have been shaped essentially by this appreciation. The field of HAM encompasses a myriad of local and global infrastructure problems. This chapter deploys and demonstrates our overall long-term architecture blueprint for a robust health identification and performance monitoring of infrastructure systems.

Taken as a whole, the characteristics of current methodologies place serious limitations on their practical use. Indeed, there are no paradigms of reliable strategies in this field that are robust enough to be of practical use. This section of the chapter will, presumably, revitalize greater and consistent acceptance of our HAM technologies. The ultimate benefit of this is an optimization of both the infrastructure performance as well as costs of repairs to extent the service lives.

Finally, it is important to recognize that although much progress has been made in this field to-date, our understanding of all technical and economic aspects is incomplete. It is anticipated that the technological vision presented herein will serve as a platform from which pertinent advances may be launched leading to more confidence in our infrastructure.

6.2 Structural health condition assessment

In general, the process of health condition assessment and performance monitoring can be categorized into five stages:

1. Identification of damage presence.
2. Localization of damage.
3. Identification of the damage type.
4. Quantification of damage severity.
5. Prediction of the remaining service life.

Customarily, the paradigms for HAM can be depicted as a four-part process:

1. Operational evaluation.
2. Experimental measurements.
3. Relevant data extraction for damage features.
4. Analytic investigation.

Operational evaluation defines why the process is to be performed, tailors its unique aspects of the system and unique features of the damage that is to be identified. The operational evaluation answers four questions regarding the implementation of a HAM system:

1. How is damage defined for the system?
2. What are the conditions under which the system to be monitored functions?
3. What are the limitations on acquiring data in the operational environment?
4. What are the economic and/or life safety motives for performing the monitoring?

Current health monitoring experimental techniques require that the vicinity of the damage is known a priori and that the portion of the structure being inspected is readily accessible. Some of the drawbacks of these methodologies include:

1. The use of expensive instrumentations and data processing hardware not designed specifically for structural health monitoring. The relative expense of these sensing technologies dictates that a structure must be sparsely instrumented

2. Varying environmental and operational conditions produce changes in the system's structural response that can be easily mistaken for damages

These drawbacks place serious limitations on the practical use of existing methodologies. Indeed, there are no HAM paradigms of reliable strategies that are robust enough to be of practical use. Our concept for the future HAM is based on a fresh look at the overall picture. Thus, we divide the future HAM into two categories:

1. Intelligent knowledge-based systems which prognosis potential damages.

2. Existing damage identification and structural performance monitoring.

This new concept emerges from an open-end modular approach that can be adapted to any system, i.e., is not limited to a particular type of damage or material, damage detection or identification techniques. Our HAM will equally perform supervised and unsupervised learning. Supervised learning refers to situations where data from damaged and undamaged structures are available. Unsupervised learning refers to situations where data is only available from the undamaged structure.

6.3 Damage assessment and monitoring

We have been tackling the problems associated with structural health identification and health monitoring. This learning process has evolved and culminated in presenting this part of the chapter. However, because there are several confounding factors and unanswered questions in this field, we embarked on the idea of importing "smartness" to the process. To adequately probe damages, we should search for characteristics such as: the existence of damages, whereabout of the damaged areas, the type of damages, the size of damages, and the remaining service life in the system.

In answering the above questions, DIM is our promising candidate for HAM utilization. Some of its unique capabilities include the following:

1.DIM's toolkits for damages identification.

2.DIM facilitates the complex HAM process.

3.DIM uses modular architecture which allows for adaptation to new development.
4.IM has a built-in database for most of the available technical information to avoid any incorrect or impractical input data.

7. DIM: geotechnical applications

Under the immense pressure of environmental, energy, economic, and other modern problems, many new materials have been developed from a wide range of renewable, natural, and manmade resources. These innovations meet urgent needs

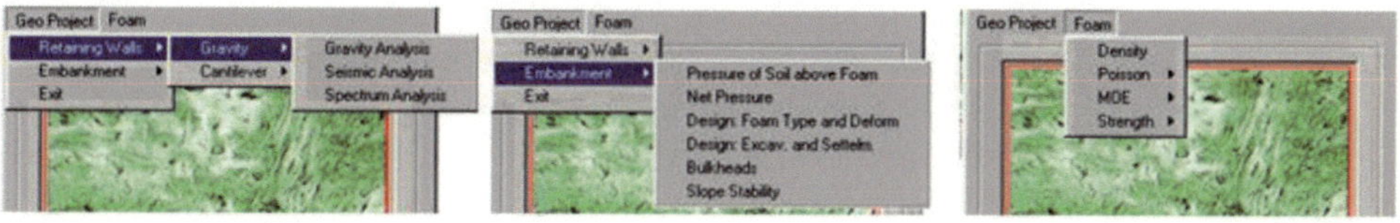

Figure 48.
Illustrative screens for the DIM' geo client.

such as high strength/weight ratio, thermal insulation, design flexibility, durability, low moisture absorption, ease and speed of construction, and suitability for adverse weather conditions. DIM incorporates a module for the applications of the emerging materials in geotechnical applications such as embankment, retaining systems, foundations (https://aitechm.wixsite.com/8kbextech). As shown in **Figure 48**, the scope covers the following:

1. Static.

2. Seismic and spectrum.

3. Slope stability.

4. Strengths.

5. Stress distributions.

6. Uplift and net stresses.

7. Designs

8. Serviceability.

7.1 DIM: biodegradable, bio-based, and synthetic load-bearing components

As elaborated previously, there is no HAM available for the newly emerging load bearing components. The DIM fills this gap and provide many capabilities for biodegradable, bio-based, and synthetic based panels and plates used as load bearing components. It should be noted that the analytical background of this part is presented earlier in the chapter (https://aitechm.wixsite.com/8kbextech). The scope of the DIM includes the following, as shown in **Figure 49**:

1. Loads:

    - Different patterns of transverse loads.

    - In-plane loads.

2. Instability:

    - Shear.

    - Uni-axial.

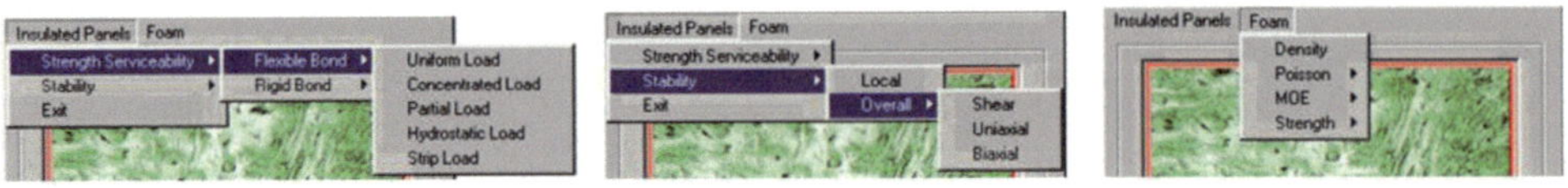

Figure 49.
Illustrative screens for of the DIM' bio load bearing client.

- Bi-axial.
- Local instability.

3. Bonding:

    - Rigid bonding.
    - Imperfect bonding.

7.2 DIM: utility network grids - poles

Overloading, aging, and environmental impacts are inevitable blows to the utility grids over the lifetime. Unfortunately, these adverse acts result in deterioration. For this specific problem, our overall goal is to develop a specific DIM with the following objectives:

- Determination if a damage is present in the pole.
- Determination of the location of the damage.
- Quantification of the severity of the damage.
- Prediction of the remaining service life of the pole.

Existing pertaining DIM methodologies rely on either visual inspection or localized measurements. Those methods include acoustic measurements and manual sample extraction techniques. Although many of these techniques found applications in the field, there remain some problems that restrict their effectiveness such as:

1. All the techniques require that the vicinity of the damage is guessed a priori and that the portion of the pole being inspected is readily accessible.
2. Contaminating the poles during extraction of test samples.
3. Soil excavation is required; thus, disturbing the supporting conditions.
4. The attachments were never considered in any of the methods.

The adverse impacts of these factors on the use of existing methods were studied by testing 30 real full-scale poles to true failure. The results indicated that none of the available methods was sufficiently accurate as a means for identifying the damages or predicting strengths of poles. Thus, the needs for a new DIM has focused on the development of a practical, fully automated, truly autonomous, human-free, inexpensive, and accurate DIM for in-situ poles.

The underlying principle of our new DIM for poles is that dynamic signatures are sensitive indicators of structural integrity. This reasoning is based on the fact that damages change the dynamic characteristics, and therefore, fundamental measurements can be used to identify and monitor damages. This approach is very attractive for the development of this DIM for poles. We believe that a knowledge-based method is the key for achieving this goal. The new DIM is posed as an inverse problem in which the analytic goal is to find pole characteristics using minimum response information obtained from in situ measurements. This inverse problem is formulated based on the rigor fundamentals of engineering mechanics and structural dynamics and is solved using an implicit redundant representation genetic algorithm. The proposed method is capable of locating and quantifying the severity of damage and remaining strength. Because the damaged parts and their locations are not known beforehand, an iterative solution process is necessary to overcome the difficulties associated with the large number of possible output scenarios. For the first time in the literature, we incorporated vital parameters such as the effects of the surrounding soils, the embedded length of poles, and the attachments. Unlike all the existing methods, our approach neither applies sophisticated numerical techniques as the finite element nor employs the traditional destructive techniques. **Figure 50** shows a selected screen for our DIM toolkit. In situ setup include recording the dynamic signal of poles, FFT analysis of the signals, and integrating the results with the presented analytic method to determine the previously mentioned characteristics of the defective zones.

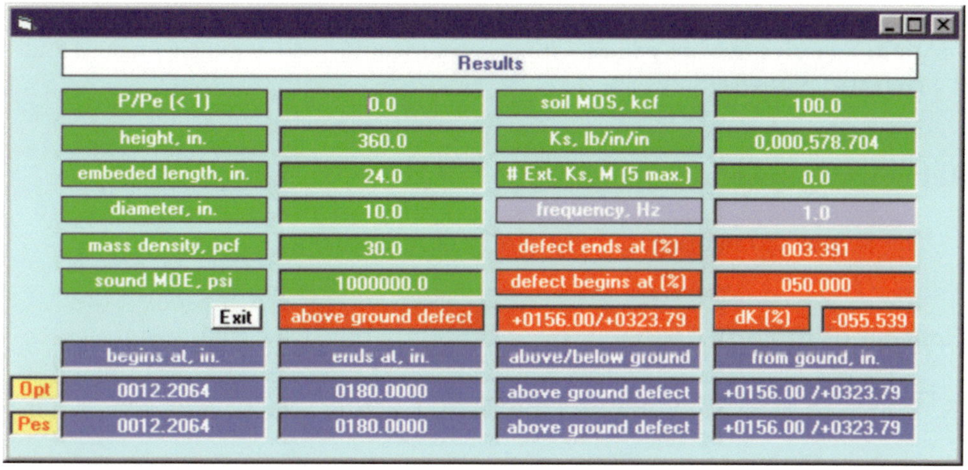

Figure 50.
Illustrative screen for the DIM' poles client.

Damage (%)		Frequency		
Begins at	**Ends at**	**FE**	**DIM**	**%**
95	100	35.9	36.3	1
90	100	23.4	23.7	1
58	70	42.6	43.0	1
Sound pole with overhead		118.3	118.2	0

Table 1.
Comparison of results obtained using the presented technology and FE.

To verify the procedure introduced in this chapter, selected problems were used in conjunction with the previous analytical method. **Table 1** compares the predicted status of selected poles and the corresponding finite element solutions. The table shows very close agreement between the results. This agreement has built confidence in the accuracy and validity of the introduce procedure for utilities (https://aitechm.wixsite.com/8kbextech).

7.3 Conclusions

The developments of modern materials and the unprecedented advances of technologies brought to the applications of green laminated composites to the forefront of construction systems. The use of adhesives is unavoidable in manufacturing the panels. However, the assumption of perfectly rigid bonding is very common in the available literature and industries. This assumption is unrealistic because adhesives like all materials have finite properties. This section in the chapter replaced the hypothetical assumption with realistic one. Closed form solutions were presented for load bearing composite panels under flexural and in-plane loads. Unlike the available theories in the literature, the solutions used bonding with realistic stiffness rather than the commonly used perfectly rigid adhesives. The solutions satisfy the equilibrium equations of each lamina, the interlayer deformation and stress compatibility equations, and the panel boundary conditions. To ascertain the effects of bonding stiffness on the performance, numerical results were obtained. The results proved that low bonding stiffness has a strong effect on the laminates deflection. For economical and safety reasons, a new parameter was introduced in the form of core-to-adhesive stiffness ratio. This ratio suggests that very stiff adhesive used in panels with soft core would yield expensive panels and the converse is unwise.

Acknowledgements

This section did not receive funds for its development.

Conflict of interest

The author declares no conflict of interest.

Author details

Rafaat Hussein
SUNY, New York, USA

*Address all correspondence to: ezpscg@gmail.com

References

[1] Ahmed M. et. al., Recent Advances in Green Composites, Materials Science Engineering, Vol. 425, 2010, pp 107–166, https://doi.org/10.4028/www.scientific.net/KEM.425.107.

[2] Bujanovic B, SUNY (2018) Personal communication

[3] Hussein, R. (2019) Artificial Intelligence for Performance and Monitoring of Sustainable Built and natural Environments, and the Climate, Chapter 7, Advances in Artificial Intelligence, the International Frequency Sensor Association.

[4] Karbhari VM (1997) Application of composite materials to the renewal of twenty-first century infrastructure. Proceedings of the Eleventh Conference on composite materials. Melbourne Australia

[5] Mohanty A et al (2005) Natural fibers, biopolymers, and biocomposites. CRC Press, Florida

[6] Rigoberto B et al (2015) Sustainable cellular biocomposites from natural fibers and unsaturated polyester resin for housing panel applications. J Polym Environ 13(2):139–149

[7] Satyanarayana KG (2007) Biodegradable composites based on lignocellulosic fibers. International Conference on advanced materials and composites (ICAMC)

[8] The University of Maine (2019) Advanced structures and composite center. http://www.aewc.umain e.edu/

[9] Vanderbilt MD, Goodman JR, Criswell ME (1974) Service and overload behavior of wood joist floor systems. J Struct Eng Div ASCE 100(ST1):11–29

[10] Placidi L et al (2018) A strain gradient variational approach to damage. A comparison with damage gradient models and numerical results. Math Mech Complex Syst 6:77–100

[11] Placidi L, Barchiesi E (2018) Energy approach to brittle fracture in strain gradient modelling. Proc R Soc Math Phys Eng Sci 474:878

[12] Placidi L et al (2018) Two-dimensional strain gradient damage modeling: a variational approach. Zeitschrift fur Angewandte Mathematik und Physik 69:56

[13] Allen HG (1969) Analysis and design of structural sandwich panels. Pergamon Press Ltd, Oxford

[14] Noor AK et al (1996) Computational models for sandwich panels and shells. Appl Mech Rev 49(3):155

[15] Plantema FJ (1966) Sandwich construction—the bending and buckling of sandwich beams. John Wiley and Sons Inc, Plates and Shells

[16] Goodman JR, Popov EP (1968) Layered beam systems with interlayer slip. J Struct Eng Div ASCE 94:2535–2548

[17] Goodman JR (1969) Layered wood systems with interlayer slip, Wood Sci 1 (3):192

[18] Goodman JR, Vanderbilt MD, Criswell ME, Bodig J (1974) Composite and two-way action in wood joist floor systems. Wood Sci, 7(1)

[19] Hussein R (1986) Composite panels/plates—analysis and design. Technomic Publishing Inc

[20] Allansson, A., "Stability and collapse of corrugated board;" MS Thesis, Lund University, 2001

[21] Hussein R (2002) Exact closed-form solutions for fiber-fiber and bond-fiber

interactions with imperfect bonding. Progress in Paper Physics Conference, Skaneateles

[22] Hussein R (2020) An analytic solution for fiber-bond interactions, The 116th ESPRA Conference, Bethlehem

[23] Nordstrand, T., "Basic testing and strength design of corrugated board and containers," Doctoral thesis, Lund University, 2003

[24] Hussein, R. (2020) Knowledge-based Tools for Monitoring and Management, and Design of the Engineered Infrastructure Construction Systems, Chapter 5, Advances in Computers and Software Engineering, the International Frequency Sensor Association.

[25] Huang J, Li K (2008) A new soy flour-based adhesive for making interior type II plywood. J Am Oil Chem Soc 85 (1):63–70

[26] Hussein, R. (2020) Knowledge-Based Expert System Technologies for the Built & Natural Environments, https://aitechm.wixsite.com/8kbextech.

Chapter 2

Construction Technology of Precast Pier Foundation Filled with Demolished Concrete Lumps

Wengui Li, Bin Lei, Zhiyu Luo and Fuzhi Yang

Abstract

Applying of demolished concrete lumps (DCLs) in the pier foundation is an effective way to improve the efficiency of construction waste resource utilization. Fifty-two cylindrical specimens with the size of ∅ 250 mm × 500 mm were fabricated by mixing of DCLs with the fresh concrete (FC) and used to investigate the influence of two key factors, the gradation of the DCLs and the height setting of layered "steel mesh," on the uniaxial compression and flexural strength properties of the compound concrete specimens. Results indicate that the layered "steel mesh" in the specimens can restrain the settlement and segregation of the DCLs and improve the compressive and flexural strength of the specimens significantly. Normally, there are two types of failure damage mode of the test pieces, the failure of the interface between DCLs and the FC and the fracture failure of the DCLs. When the stress level is below 0.5, the test piece is in the elastic stage. Crack development occurs when stress level further increase to 0.7–0.9. The pieces with the layered pouring height of H2 and the DCLs of R3 present the optimum compressive strength and flexural strength and also best construction effect.

Keywords: demolished concrete lump, precast pier foundation, construction technology, uniaxial compression test, flexural test

1. Introduction

Currently, China is in an important stage of urbanization development. A large number of construction wastes are generated in a series of projects such as the large-scale reconstruction of old cities and demolition of temporary or illegal buildings. It has become a trend to accelerate the resource utilization of construction waste. Literature [1] proposed a new method of recycling of waste concrete components, which use the large size of waste concrete block directly in the member. Instead of the normal way of using the waste concrete aggregate (fine aggregate size of 0.075–4.75 mm and coarse aggregate size of 4.75–40 mm) [2, 3], this method reuses the construction waste at the large block scale or even segmental level (scale 40 mm or higher), which thereby avoids the cumbersome process of production of recycled aggregate and results in a significant reduction of cost. The material obtained by mixing the waste concrete block with the new concrete is called as the regenerated block concrete. A large number of relevant studies have been conducted, which mainly include strength prediction of the compound concrete made of demolished concrete lumps (DCLs) and fresh concrete

(FC) [4–8], failure characteristics [9, 10], size effect [10, 11], basic creep [12–14], freeze–thaw mechanical properties [15, 16], pouring process [7, 17–19], hydration temperature rise [20, 21], impermeability [21], etc. Abundant theoretical achievements have been made based on these experimental studies. For field application, the exploration on the application of this compound concrete mainly focuses on the steel-concrete composite structure. There are good theoretical basis and broad prospect for the engineering application of the compound concrete [18].

Pier foundation is a foundation form between the rigid independent foundation and the manually dug pile foundation [22, 23]. Combining with the favorable geological conditions of Nanchang city, it is of great value to carry out the research on pier foundation, which will have a good application prospect in the foundation and foundation reinforcement treatment of multi-storey houses (such as villas and houses). Due to the large volume of the pier foundation, the large-diameter DCLs can be directly poured into the pier foundation with the FC in an appropriate ratio, thereby improving the efficiency of building resource utilization. At the same time, under the background of green, energy-saving, environmental protection and sustainable development as well as from the motivation of simplifying the material consumption and energy consumption for recycling of DCLs, the concept of precast pier foundation with DCLs mixed with FC is put forward. By the method of prefabrication, the quality of pier foundation filled with DCLs and FC can be well guaranteed. At the same time, it can achieve benefits such as more convenient construction, no noise [24], and construction cost-efficiency. This method is especially helpful when there is no favorable condition for site construction, the pump truck cannot arrive at the scene, there is a need to speed up the construction progress, etc. Research on precast pier foundation filled with DCLs and FC of the research has very important value.

In order to apply the compound concrete better in the pier foundation, the experimental research on the construction technology of the precast pier foundation filled with DCLs and FC was carried out. For related literatures mentioned above, it was found that current experimental studies for the compound concrete mainly concentrate on several factors including size of DCLs, mass replacement rate, strength difference of waste concrete strength, different proportions of FC and the shape of test piece shape, etc. However, there is no research for the gradation of DCLs, and the engineering problem of concrete segregation is likely to occur during the pouring and vibration process. In this paper, the corresponding test plan is proposed for these problems, and the suggestions for guiding the construction of precast pier foundation of compound concrete are obtained through test analysis. It provides theoretical basis for the engineering application of precast pier foundation filled with compound concrete.

2. Experimental procedures

2.1 Raw materials

The component materials of the compound concrete are the following: (1) Cement: Nanchang conch brand with 42.5 grade ordinary Portland cement (P.O 42.5) is adopted. The initial setting time of the cement is 203 min, and the final setting time is 250 min. (2) Fine aggregate: Ganjiang sand with apparent density of 2468.4 kg/m^3 and particle size less than 4.75 mm is adopted. It is a medium sand and is graded in zone II. (3) Coarse aggregate: crushed stone was used as coarse aggregate in the test, with particle size range 4.75–20 mm. (4) Water: the test water used is the potable tap water. (5) DCLs: in the test, the DCLs were taken from the concrete removed from the back street of Nanchang University and then processed into DCLs with particle size of 40–80 mm by crushing tools. According to the needs of this research, the

Figure 1.
Demolition and transformation of DCLs.

Figure 2.
Test DCLs after treatments.

original DCLs were sorted into two piles of DCLs with particle size of 40–60 mm and 60–80 mm, respectively, as shown in **Figures 1** and **2**. Before pouring, the DCLs must be wet, and the saturated surface-dry aggregates were used for pouring. (6) Fly ash: Hougang grade I fly ash was used in the test. The main components of fly ash are SiO_2 and Al_2O_3, accounting for more than 60% of the total content. (7) High-efficiency water-reducing agent: SM-F ultra-dry powder active high-efficiency water-reducing agent was used in the test, and the commonly used dosage was 0.5–1.5% of the cementitious materials.

2.2 Specimen design and production

A total of 52 cylinder specimens with size of ∅250 mm × 500 mm are used for this research, which consist of 13 groups of specimens with 4 specimens for each group. Research involves the following two factors: (1) The grade of DCLs. Four kinds of gradation with 40–60 mm to 60–80 mm DCL ratios of 2:8, 4:6, 6:4, and 8:2 are designed [25]. (2) The setting of the layered barrier "steel mesh." The design

w/c	Water (kg/m^3)	Cement (kg/m^3)	Coarse aggregate (kg/m^3)	Fine aggregate (kg/m^3)	Fly ash (kg/m^3)	Water reducer (kg/m^3)
0.49	190	282	1087	688	98	6.46

Table 1.
Mix proportions of fresh concrete.

Notation	Number of specimens	Height of the layered pouring	Particle size grading of DCLs (40–60 mm: 60–80 mm)	$f_{cu,new}$ (MPa)	$f_{cu,old}$ (MPa)	η (%)
H1-R0	4	0 mm layer	—	36.37	22.12	30
H1-R1	4		2:8			
H1-R2	4		4:6			
H1-R3	4		6:4			
H1-R4	4		8:2			
H2-R1	4	100 mm layer	2:8			
H2-R2	4		4:6			
H2-R3	4		6:4			
H2-R4	4		8:2			
H3-R1	4	200 mm layer	2:8			
H3-R2	4		4:6			
H3-R3	4		6:4			
H3-R4	4		8:2			

Note: $f_{cu,new}$ and $f_{cu,old}$ represent the compressive strength of DCLs and FC, respectively. H represents the height of layered pouring, and R represents the gradation of waste concrete blocks.

Table 2.
Specimen parameters.

considered three kinds of layered method, which are 0, 100, and 200 mm layer. The quality of the construction technology are reflected and estimated by the mechanical properties of the specimens.

Specimens were made by mixing of DCLs with FC, and the same mixture ratio with water to cement ratio (w/c) of 0.49 is adopted for FC as given in **Table 1**. The compressive strength value of the FC test cube is 36.37 MPa. According to the results from rebound hammer, the compressive strength value of the DCLs is determined as 22.12 MPa. The replacement ratio η of the DCLs is set as the constant value of 30% [10], which is defined as the mass proportion of the DCLs to the overall specimen. The samples designed with different parameters are summarized in **Table 2**. The manufacturing process of compound concrete specimen is shown in **Figure 3**.

2.3 Test loading scheme

2.3.1 Uniaxial compression test of the specimen

Microcomputer-controlled servo universal testing machine (model: SH-T4306) in Nanchang University mechanics laboratory is used for uniaxial compression test. Four YWC-30 type displacement meters with the range of 30 mm and accuracy of 0.005 mm are set around the specimen with a space of 90° to measure the displacement changes in the process of compression. The displacement meters

Step 1: pour 30 mm FC and then place the wire mesh

Step 2: put in the prepared DCLs

Step 3: add FC with vibration

Step 4: when reaching 37 mm, vibrate until there is no obvious bubble and the surface is flooded

Step 5: place in the wire mesh for the second time

Step 6: put in the prepared DCLs for the second time

Step 7: repeat the above steps until the mould is filled

Step 8: vibrate the concrete to compact and smooth the surface

Step 9: formed specimen

Figure 3.
Construction process of specimens.

are placed in the middle of the specimen and the measuring distance is 0.4 times the height of the specimens. During the test of cylindrical specimens, the pressure direction was consistent with the pouring direction. The flatness of the upper surface of casted cylindrical specimens could not easily meet the compression test requirements. Therefore, 1 cm standard fine sand was laid on the upper surface of uneven specimens during the test. In order to prevent the sand laid on the upper surface of the specimen falling into the oil cylinder and damaging the press equipment, the specimen is firstly put into a stainless steel basin and then placed on the press machine. According to ASTM C39 (Standard Test Method for Compressive Strength of Cylindrical Concrete Specimens) [26], the loading rate was set as 0.3 mm/min, and the test data were collected using UT7116Y static strain gauge with a collection frequency of 1 HZ. Before loading the press machine, the pre-loading value is set at about 30% of the peak load. The test device is shown in **Figure 4**.

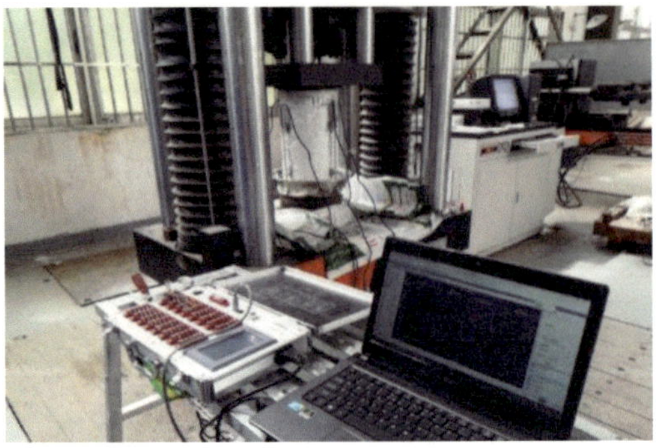

Figure 4.
Uniaxial compression test device.

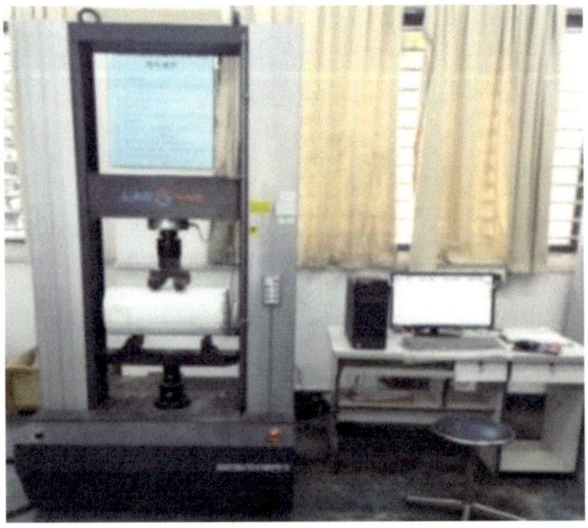

Figure 5.
Flexural strength test equipment.

2.3.2 Flexural strength test of specimens

Microcomputer-controlled universal electronic testing machine (model: LD26.305) in the structure laboratory of Nanchang university is adopted for the flexural strength test. The range of this machine is up to 300KN and the accuracy grade is 0.5. For the size of ∅250 mm × 500 mm of the casted specimens in this research, it does not meet the requirements for the flexural strength test. However, according to the bending test study for rectangular pavement brick with different spans between two supports [27], when the cross/thickness ratio is 2:1 or 3:1, there is no significant influence of the support span on the flexural strength of the rectangular road brick. As a result, it is feasible to use this kind of sample for bending test. According to the provisions of 4.6.3 in the code [28], the specimen is preloaded. The preloading load value is equal to 20% of the peak load, and the load is continuously and uniformly applied at the rate of 250 N/S. The test device is shown in **Figure 5**.

3. Uniaxial compression

3.1 Experimental phenomena

In this study, uniaxial compression tests were carried out on 13 groups of 26 specimens, and the results showed that no cracks were observed on the surface of the

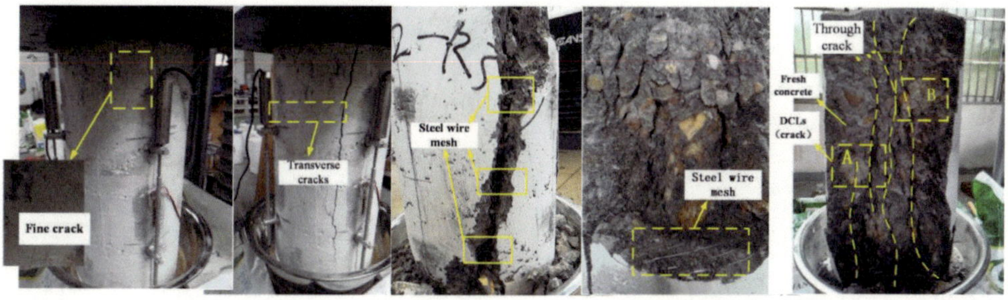

(a1) cracking load (a2) peak load (a3) failure (a4) internal damage
(a)
(b1) internal damage
(b)

Figure 6.
Failure modes of typical samples. (a) Failure of H2 and H3 in various loading stages, (b) Failure of H1.

specimens at the initial stage of loading. Besides, when the load increases continuously to about 70–90% of the peak loading, tiny cracks start to appear at both ends of the specimen, as shown in **Figure 6(a1)**. Further increasing the load to the peak load leads to the gradual expansion of the micro-cracks on the surface of samples, and the place where the "steel wire mesh" is placed generates cracks around the specimen. With the increase of the load, the crack propagation on the surface of specimen is relatively slow, and only few penetrative cracks are formed, but the crack ratio will gradually increase. After achieving the peak load, the specimen damage slows down with the loading. When the load decreased to 50–65% of P_u (P_u represents peak load), the load remains almost unchanged with the increases of displacement. At this period of time, the "wire meshes" restrain the rapid crack of specimens, improve the ductility of the specimens, and also improve the compressive strength of recycled concrete block, as shown in **Figure 6(a2)** and **(a3)**.

As revealed in **Figure 6(a3)** and **(a4)**, when constrained by "wire mesh" for crack, specimen maintains its integrity. At the same time, the "wire meshes" in the specimen also help to decrease the settlement of DCLs in the process of casting the samples, making it distribute uniformly in the specimen. In **Figure 6 (b)**, it can be seen that the interface between the DCLs and the FC is well poured without any cracks. By observing areas A and B in the figure, it can be found that the DCL is cracked, while the interface between the DCLs and the FC is not cracked, indicating that the DCLs and FC interface are not the obvious weak parts of the specimen.

3.2 Analysis of test results

The top surface treatment method and test values of each specimen are given in **Table 3**. Compared with the test results in literature [5, 10], the test results in this study are relatively low. The main reason is attributed to the absence of transverse constraint effect, namely, the "hoop effect" on the test piece due to the addition of sand on the uneven test surface. Besides, the size effect of the test piece also exacerbates this phenomenon.

3.3 Influence of gradation of DCLs on compressive strength

Based on the compressive strength values of each specimen in **Table 3**, the change of compressive strength with the gradations of DCLs for H1, H2, and H3 specimens classified by layered pouring height is given in **Figure 7**. According to the trend in **Figure 7**, the following conclusions can be drawn: (1) the compressive strength of compound concrete increases with the increase of the proportion of DCLs with

Notation	Surface	Compressive strength (MPa)		Notation	Surface	Compressive strength (MPa)	
		Test results	Average value			Test results	Average value
H1-R0–1	Place sand	15.77	15.95	H2-R3–1	Place sand	12.71	12.02
H1-R0–2	Place sand	16.12		H2-R3–2	Place sand	11.33	
H1-R1–1	Place sand	9.93	9.93	H2-R4–1	Place sand	11.49	11.38
H1-R1–2	No sand	14.13		H2-R4–2	Place sand	11.27	
H1-R2–1	Place sand	11.27	11.02	H3-R1–1	Place sand	10.72	11.38
H1-R2–2	Place sand	10.77		H3-R1–2	Place sand	12.04	
H1-R3–1	Place sand	12.52	12.37	H3-R2–1	Place sand	12.21	11.67
H1-R3–2	Place sand	12.21		H3-R2–2	Place sand	11.12	
H1-R4–1	Place sand	11.07	11.26	H3-R3–1	Place sand	12.34	13.54
H1-R4–2	Place sand	12.04		H3-R3–2	Place sand	14.75	
H2-R1–1	Place sand	11.73	11.73	H3-R4–1	No sand	14.28	11.51
H2-R1–2	No sand	15.32		H3-R4–2	Place sand	11.51	
H2-R2–1	Place sand	10.11	11.86				
H2-R2–2	Place sand	13.61					

Note: For average compressive strength of H1-R1, H2-R1, and H-R4, the compressive strength is taken directly from the sample with sand.

Table 3.
Results of compressive strength test.

particle size of 40–60 mm until it reaches 60%. When exceeding 60%, further increase in the content of DCLs with particle size of 40–60 mm would lead to strength reduction. (2) With the change of DCL grade, the specimens H1, H2, and H3 all show the maximum compressive strength when the grade is R3, which corresponds to the case that the ratio of 40–60 mm DCLs to 60–80 mm DCLs is 6:4. It is caused by a reason that under this gradation, the bulk density of DCLs is the highest, namely, a better natural gradation [29]. (3) The gradation of DCLs has a certain impact on the strength of compound concrete. For H1-type specimens, the compressive strength of H1-R3 is 24.57% higher than that of H1-R1. Similarly, the compressive strength of H2-R3 is 5.62% higher than that of H2-R4, and the compressive strength of H3-R3 is 18.98% higher than that of H3-R1. (4) With the same proportion, the strength of compound concrete specimens mixed by both FC and DCLs is lower than the specimens just made of FC. The greatest reduction occurs for H1-R1 sample. Compared with H1-R0, the compressive strength of it decreases by 37.7%.

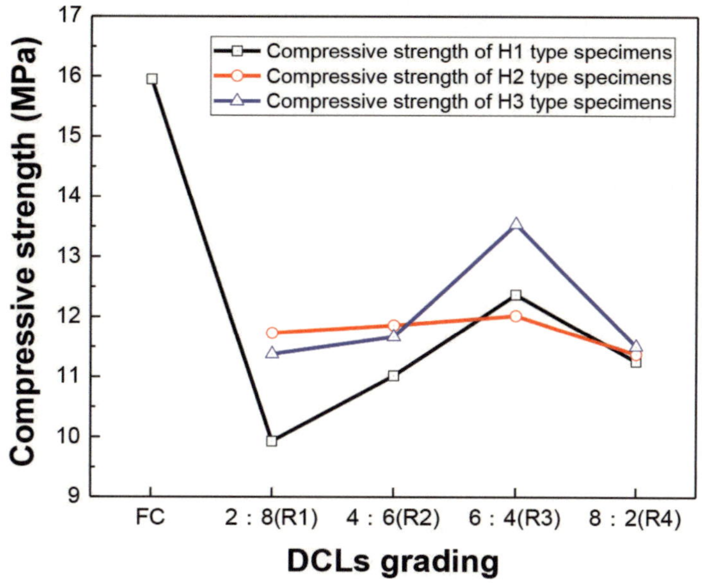

Figure 7.
The compressive strength of compound concrete with different DCL gradations.

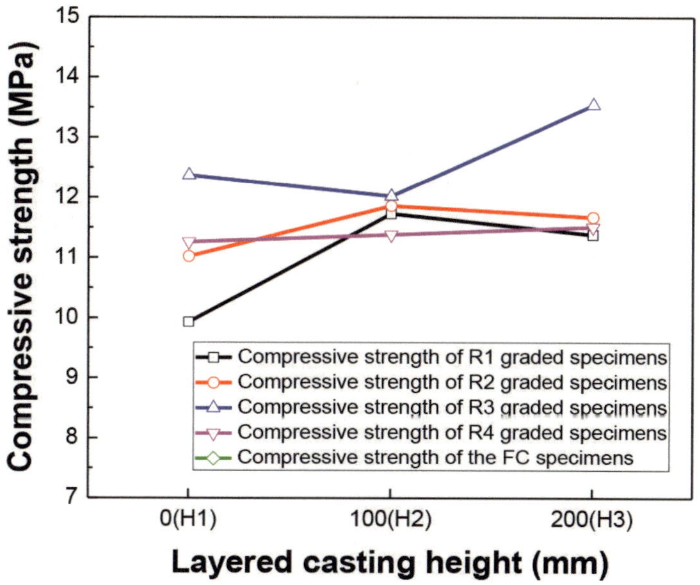

Figure 8.
The compressive strength of compound concrete with different layered casting heights.

3.4 Influence of layered pouring height on compressive strength

According to the compressive strength values of each specimen in **Table 3**, the influence of different layered pouring heights on the compressive strength of classified samples R1, R2, R3, and R4 with different DCL gradations is shown in **Figure 8**. The following conclusions can be drawn according to the results in **Figure 8**: (1) the compressive strength of compound concrete increases by 1.05–12.27% after steel wire mesh is put into each layer when layered pouring is adopted for R1, R2, and R4 gradations. (2) With the increase of the height of the layered pouring, the overall performance change is not the same for all samples. The compressive strength of samples with R1 and R2 gradation DCLs increases first and then decreases with the increase of layered pouring height, which may be because the "steel wire mesh" is more in layered height of H2 specimen than in H3 specimen. The "steel wire mesh" inhibits the transverse deformation of the specimen, consumes part of the energy, and improves the compressive strength of the specimen. The compressive strength of sample with R4 gradation DCLs remains stable, indicating that this gradation is reasonable and plays a major role in the strength of compound concrete. The compressive strength of sample with R3 gradation DCLs is decreased first and then increased with the increase of the layered pouring height. The compressive strength value of this sample is higher than that of other gradation specimens when has the same layered pouring height. (3) When the height of layered pouring is set as H2, the measured compressive strength is more stable and less discrete.

3.5 Analysis of stress: strain relationship curve

Studying the changing trend of stress–strain curve is an important way to understand the deformation properties of specimens. In this paper, 13 groups of tests were carried out. Based on the analysis and consideration of both the two factors, it can be seen that the casting quality of the specimens is better when the gradations are R3 and R4, and it is easier to ensure the casting quality when the height of the layered casting is H2. Therefore, the stress–strain curves of R3 and H2 specimens were selected for comparative analysis in this paper to provide the theoretical basis for the application of compound concrete in prefabricated pier foundation.

As shown in **Figures 9** and **10**, the following conclusions can be drawn from the analysis: (1) the slope of the origin tangent line of the stress–strain curve of the FC specimen is greater than that of the specimen with DCLs, and the development trend of the stress–strain curve of the specimen with DCLs is consistent with that of the FC specimen. (2) The greater the compressive strength of the specimen in the steeper rising section of the curve, the greater the stiffness of the specimen. (3) Increasing the proportion of DCLs with particle size ranging from 40 to 60 mm

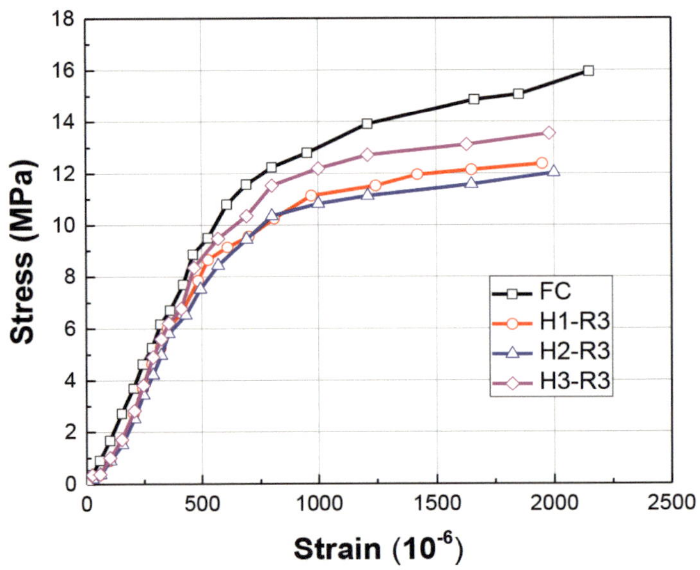

Figure 9.
Comparison of stress–strain curves of R3 specimens.

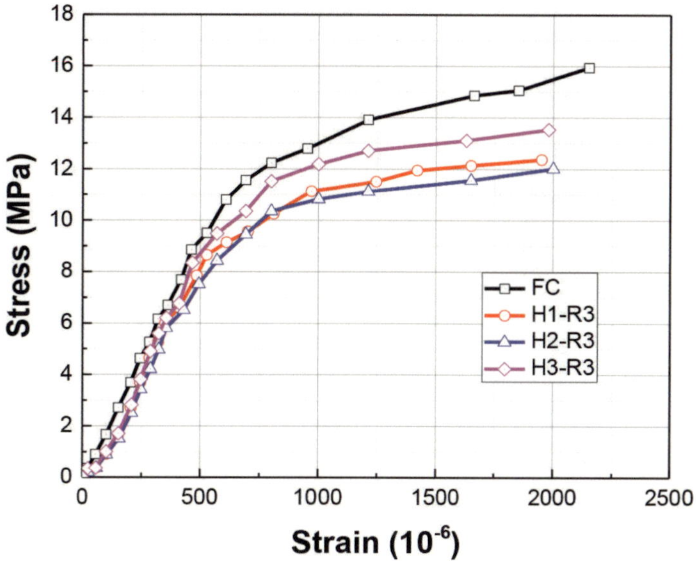

Figure 10.
Comparison of stress–strain curves of H2 specimens.

leads to the steeper rising section of the stress–strain curve and the smaller peak strain. When the particle size exceeds 60%, with the increase of the content of corresponding DCLs, the rising section of the stress–strain curve becomes gentle, and the peak strain increases. (4) The peak strain of the compound concrete specimen is in the range of 1800×10^{-6} to 2200×10^{-6}. The specimens are in elastic stage when the stress is around $0.5 f_c^0$ fc0 (f_c^0fc0 refers to the peak stress) and in crack development stage when the stress reaches 0.7–0.9 fc0 f_c^0.

4. Results and analysis of flexural strength

4.1 Experimental phenomena

In this study, 13 groups and 26 pieces of specimens were tested. The results show that:

1. In the loading process, micro-cracks are found in the tensile area of the pure bending section, which is, at the bottom of the pure bending section. With the increase of the load, the crack of the specimen gradually develops toward the compression region, and the final typical failure form of the specimen is shown in **Figure 11**. The failure crack of the specimen in the pure bending section is not through the whole section, which is affected by the steel wire mesh added into the specimen. The ultimate failure of the crack is parallel to the direction of stress, that is, the included angle is about 90^0.

2. As shown in **Figure 12**, two types of damages are identified for the fracture surface:

   a. In terms of the interfaces between the DCLs and FC as shown in **Figure 12**, the DCLs in the areas A and B correspond to the groove in areas A and B, indicating that the damage fracture of the specimen in bending test is due to the weak bond between the DLCs and FC interface.

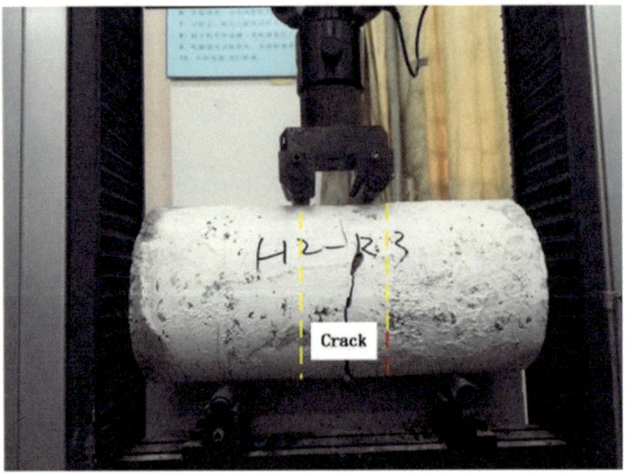

Figure 11.
Typical flexural failure modes.

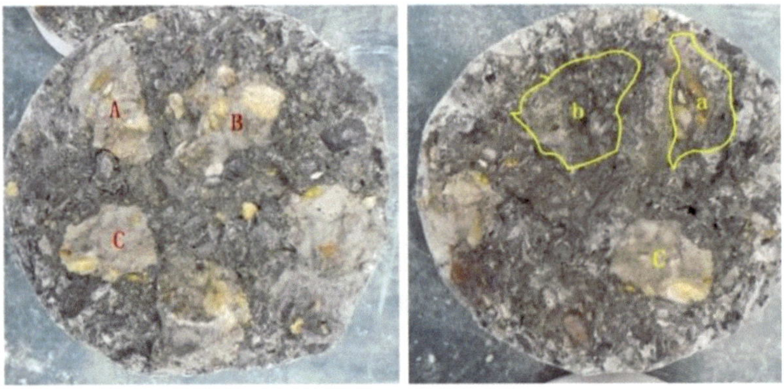

Figure 12.
Typical fracture surface.

b. The fracture failure of DCLs. Micro-cracks are generated on the DCLs during the crushing process. Besides, the strength of the DCLs is lower than that of the FC. Those factors lead to the fracture of the DCLs during flexural test.

4.2 Influence of gradation of waste concrete blocks on flexural strength

Based on the flexural strength results of each specimen in **Table 4**, the variation of flexural strength of specimens H1, H2, and H3 classified according to the height of layered pouring with the gradation of DCLs is present in **Figure 13**. The following conclusions can be drawn from **Figure 13**: (1) flexural strength of the fresh concrete specimen is higher than that of the recycled concrete specimen. Flexural strength of compound concrete arises with the increase of content of DCLs of the particle size from 40 to 60 mm. However, the limitation content for 40–60 mm DCLs is 60%, and the flexural strength of compound concrete would decrease if there is a further increase in its content. (2) As the gradation of DCL changes, the specimens H1, H2, and H3 classified according to the height of layered pouring all show the best flexural strength when the gradation is R3, namely, the ratio of 40–60 mm to 60–80 mm DCLs is 6:4. The flexural strength of H3-R3 specimen was

Notation	Flexural strength (MPa)		Notation	Flexural strength (MPa)	
	Test results	Average value		Test results	Average value
H1-R0–3	3.43	3.62	H2-R3–3	3.13	3.18
H1-R0–4	3.82		H2-R3–4	3.23	
H1-R1–3	2.84	2.81	H2-R4–3	2.92	2.96
H1-R1–4	2.78		H2-R4–4	3.00	
H1-R2–3	2.76	2.81	H3-R1–3	2.78	2.80
H1-R2–4	2.86		H3-R1–4	2.81	
H1-R3–3	2.99	3.04	H3-R2–3	2.91	2.91
H1-R3–4	3.09		H3-R2–4	2.92	
H1-R4–3	3.30	3.01	H3-R3–3	3.18	3.20
H1-R4–4	3.38		H3-R3–4	3.23	
H2-R1–3	2.88	2.92	H3-R4–3	2.83	2.86
H2-R1–4	2.95		H3-R4–4	2.88	
H2-R2–3	2.99	3.02	—	—	—
H2-R2–4	3.06		—	—	—

Table 4.
Results of flexural strength test.

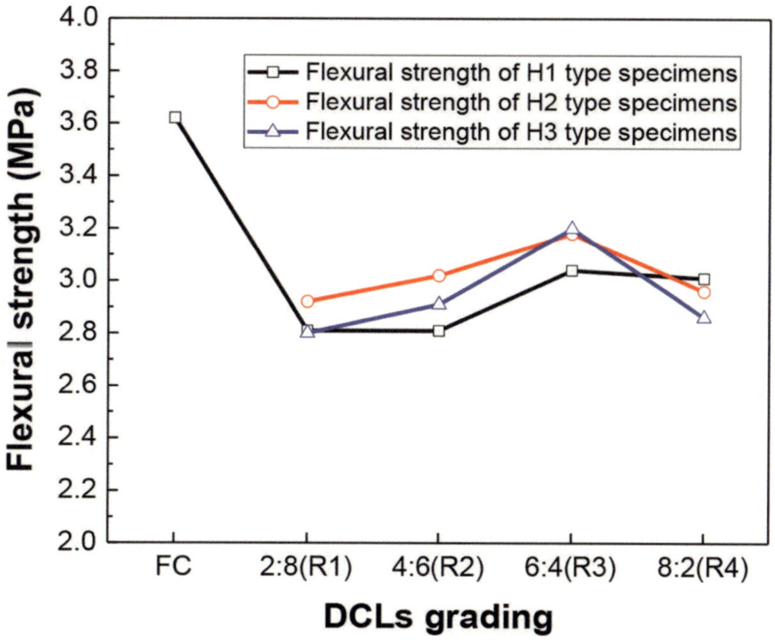

Figure 13.
Flexural strength of compound concrete with different gradations of DCLs.

14.28% higher than that of H3-R1. (3) Flexural strength of H2 type specimens is generally more regular than that of H1 and H3 type specimens. Except for differences for few situations, for instance, flexural strength of H2-R3 is 0.63% lower than that of H3-R3 and flexural strength of H2-R4 is 1.66% lower than that of H1-R4, the H2 type specimens generally show the best flexural strength.

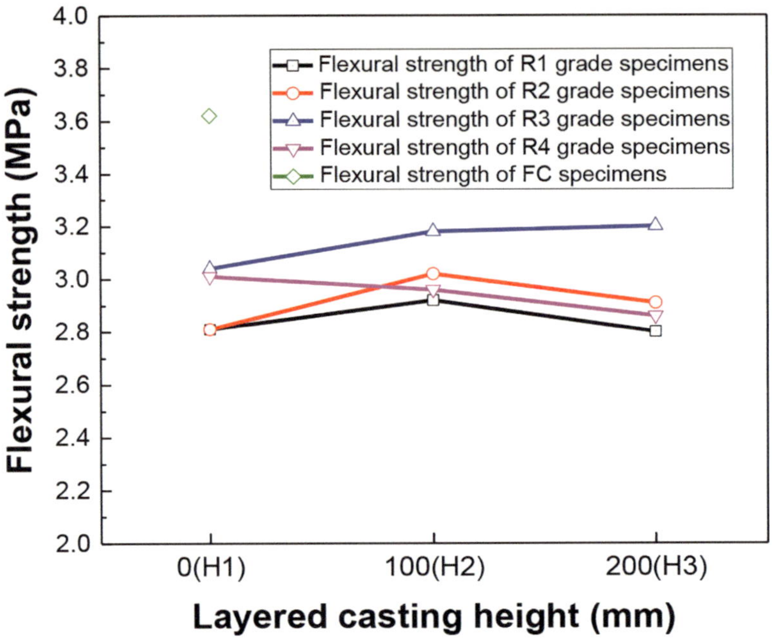

Figure 14.
Flexural strength of compound concrete with different layered casting heights.

4.3 Influence of layered pouring height on flexural strength

The influence of layered pouring height on flexural strength of compound concrete with R1, R2, R3, and R4 grade DCLs is displayed in **Figure 14**. The following conclusions can be drawn from **Figure 14**: (1) the flexural strength of compound concrete with R3 gradation DCLs is higher than that of compound concrete with R1, R2, and R4 gradation DCLs. The R1 and R2 graded compound concrete achieve their maximum flexural strength value when the lamination height of them is 100 mm, which is not the same for R3 and R4 graded compound concrete. It is hard to reveal the change law of the flexural strength as it is affected by many factors. (2) The flexural strength of the compound concrete with R1, R2, and R3 gradations of DCLs are improved with the addition of the layered steel wire mesh. However, with the addition of steel wire mesh, the flexural strength of the compound concrete with R4 gradation of DCLs decreases, which may be due to the poor casting quality of the samples under this grade, resulting in the reduction of the flexural strength.

5. Suggestions on construction technology of precast pier foundation

Based on the phenomenon and results of uniaxial compression test and flexural test, suggestions are provided for the construction of precast pier foundation filled with DCLs and FC in practical engineering. In the uniaxial compressive test, the specimens are layered pouring according to the height of "steel wire mesh," which can increase the compressive strength of the compound concrete specimens by 12.27% at most. Adjusting the gradation of DCLs can realize a maximum improvement of 24.57% for the compressive strength of the compound concrete specimens. In the flexural test, the flexural strength of compound concrete is also improved with the adjustment of the above two factors. Therefore, both the gradation of DCLs and the setting of layered "steel wire mesh" at suitable height can effectively improve the casting quality and construction technology for the construction of precast pier foundation filled with DCLs and FC. Combining the results of uniaxial compression strength test with those of flexural strength test, the specimen can achieve the optimum strength performance when the gradation of DCLs in it is R3.

According to the results of uniaxial compression test and flexural test, the specimens with layered pouring height of H2 can reach the optimum strength value. Namely, "steel wire mesh" is set according to the height of 100 mm for layered pouring. In the test, when the ratio of the diameter to the height of the specimen is set as 1:2 and the pouring layer ratio of the specimen is set as 1:5, better casting quality is achieved. In order to serve as a reference for the construction technology of the precast pier foundation filled with DCLs and FC, the concept of "pouring layer ratio" is proposed which refers to the ratio of the height of the pouring layer to the height of the specimen. According to the size of precast pier foundation filled with DCLs and FC, the height of layered pouring is set according to the "pouring layer ratio."

There are important points that should be noted during the pouring process of compound block: (1) the DCLs should be wetted before pouring to prevent the DCLs from absorbing too much water during pouring. Otherwise, it would affect the workability of FC and reduce the construction quality. (2) The sample should be vibrated by handheld vibrator during pouring. According to the code [30], there are three points to be noted as follows:

1. The insertion depth in the vibratory process should be controlled within the range of 50–100 mm. Besides, it cannot be too close to the mold, and suitable distance is around 50–100 mm. The other side should not collide with the steel cage to avoid damaging the strain gauge and affecting the measurement of test data.

2. The vibration time of each vibration point is determined by satisfying aspects that the FC would no longer sink (the surface appear floating slurry, the DCLs being surrounded by FC, and no over vibration and leakage vibration).

3. After completing a vibration process and needing to exchange the vibrate position, the vibrator has to be raised vertically slowly instead of being dragged horizontally in concrete.

The coarse aggregate particle size of the FC used in the casting of compound concrete should be no more than 20 mm, and the slump of FC should be at large value of about 160 mm. In the pouring process, the FC and the DCLs are put into the concrete repeatedly and alternately, which should be accompanied by sufficient vibration.

6. Conclusion

1. The failure phenomenon of the specimen in the uniaxial compression test shows that adding "steel wire mesh" to the specimen for layered pouring can restrain the settlement and segregation of the DCLs. In addition, during the unloading stage, the added "steel wire mesh" inhibits the rapid cracking and improves the ductility of the specimen and thus enhances the compressive strength. In the flexural test, two failure forms are present on the failure surface of the specimen, namely, the failure of interface between the DCLs and the FC and the fracture failure of the DCLs.

2. The compressive and flexural strength of the compound concrete would increase gradually with the increase of the proportion of 40–60 mm DCLs until it reaches 60%. The maximum strength achieved at the gradation of R3 (40–60 mm DCLs: 60–80 mm DCLs = 6:4) and further increase in the proportion of 40–60 mm DCLs would results in decrease of strength. The compressive strength of H1-R3 specimen was increased by 24.57% compared with H1-R1. The flexural strength of H3-R3 specimen was 14.28% higher than that of H3-R1.

3. Stress–strain curve of the compound concrete filled with DCLs and FC was analyzed. It is concluded that the slope of the origin tangent line of the stress–strain curve of the FC specimen is greater than that of the specimen with DCLs and the development trend of the stress–strain curve of the specimen with DCLs is consistent with that of the FC specimen. The peak strain of the compound concrete specimen is in the range of 1800×10^{-6} to 2200×10^{-6}. The specimens are in elastic stage when the stress is around $0.5 f_c^0$ (f_c^0 refers to the peak stress) and in crack development stage when the stress reaches $0.7–0.9 f_c^0$.

4. It is concluded from the analysis that the compressive strength and flexural strength of the specimens are best when the height of layered pouring is H2 and the grade of DCLs is R3, and it also reflects that the construction process has the best effect under this condition. The suggestions for the construction technology of precast pier foundation filled with DCLs and FC are proposed.

Acknowledgements

This work was supported by the National Natural Science Foundation of China (51668045 and 51808133) and Jiangxi Science and Technology Committee (20161BBG70056). The authors also acknowledge the research supports from the Australian Research Council (ARC), Australia.

Author details

Wengui Li[1]*, Bin Lei[1,2], Zhiyu Luo[1] and Fuzhi Yang[2]

1 Center for Green Technology, School of Civil and Environmental Engineering, University of Technology Sydney, NSW, Australia

2 School of Civil Engineering and Architecture, Nanchang University, Nanchang, Jiangxi, P.R. China

*Address all correspondence to: wengui.li@uts.edu.au

References

[1] Wu B, Liu Q, Liu W. Primary study on recycled-concrete-segment filled steel tubular members. Earthquake Resistant Engineering and Retrofitting. 2008;**4**:120-124. (in Chinese)

[2] GB/T 25176. Recycled Fine Aggregate for Concrete and Mortar. Beijing: China Building Industry Press; 2010. (in Chinese)

[3] GB/T 25177. Recycled Coarse Aggregate for Concrete. Beijing: China Building Industry Press; 2010. (in Chinese)

[4] Wu B, Zhang S, Yang Y. Compressive behaviors of cubes and cylinders made of normal-strength demolished concrete blocks and high-strength fresh concrete. Construction and Building Materials. 2015;**78**:342-353

[5] Zhang J. Experimental Study on Axial and Seismic Behaviors of Square Thin-Walled Steel Tubular Columns Filled with Demolished Concrete Lumps. Guangzhou: South China University of Technology; 2011. (in Chinese)

[6] Zhang S. Compressive Behaviors of Cubic and Cylindrical Specimens Made of Normal-Strength Demolished Concrete Blocks and High-Strength Fresh Concrete. Guangzhou: South China University of Technology; 2014. (in Chinese)

[7] Liu L. Test Study on Casting Technology and Uniaxial Compressive Behaviors of Compound Concrete Made of Demolished Concrete Lumps and Fresh Concrete. Guangzhou: South China University of Technology; 2017. (in Chinese)

[8] Wu B, Liu L, Zhao X. Test study on uniaxial compressive behaviors of compound concrete made of normal-strength demolished concrete lumps and high-strength self-compacting concrete. Journal of Building Structures. 2016;**37**(S2):73-78. (in Chinese)

[9] Wu B, Liu C, Wu Y. Compressive behaviors of cylindrical concrete specimens made of demolished concrete blocks and fresh concrete. Construction and Building Materials. 2014;**53**:118-130

[10] Liu C. Size Effect on Compressive Behaviors of Cubes and Cylinders Made of DCBs & FC and Seismic Behaviors of Double Thin Skin Hybrid Walls Filled with DCBs & FC. Guangzhou: South China University of Technology; 2013. (in Chinese)

[11] Wu B, Liu C, Yang Y. Size effect on compressive behaviours of normal-strength concrete cubes made from demolished concrete blocks and fresh concrete. Magazine of Concrete Research. 2013;**65**(19):1155-1167

[12] Yan H. Creep Behaviour of Cylinders Made of DCBs & FC and Thin-Walled Circular Steel Stub Columns Filled with DCBs & FC. Guangzhou: South China University of Technology; 2014. (in Chinese)

[13] Ai W. Creep Behaviour of Cylinders Made of Normal-Strength DCLs and High-Strength FC and Stub Steel Columns Filled with Normal-Strength DCLs and High-Strength FC. Guangzhou: South China University of Technology; 2017. (in Chinese)

[14] Wu B, Ai W, Zhao X. Test on basic creep behaviors of compound concrete made of normal-strength demolished concrete lumps and high-strength fresh concrete. Journal of Building Structures. 2016;**37**(S2):109-114

[15] Li Z. Mechanical Properties of Compound Concrete Containing Demolished Concrete Lumps at Room

Temperature and after Freeze-Thaw Cycles. Guangzhou: South China University of Technology; 2018. (in Chinese)

[16] Wu B, Li Z. Mechanical properties of compound concrete containing demolished concrete lumps after freeze-thaw cycles. Construction and Building Materials. 2017;**155**:187-199

[17] Wu B, Ji M, Zhao X. State-of-the-art of recycled mixed concrete (RMC) and composite structural members made of RMC. Engineering Mechanical. 2016;**33**(1):1-10. (in Chinese)

[18] Wang L, Jin H. Research and discussion on recycled concrete engineering application. Building Construction. 2016;**12**:1754-1756. (in Chinese)

[19] Xu J. Preliminary Application on the Engineering Adaptability of Concrete Structural Members Filled with Demolished Concrete Lumps. Guangzhou: South China University of Technology; 2012. (in Chinese)

[20] Wu B, Wu D. Test study on hydration temperature of compound concrete made of demolished concrete lumps and fresh concrete. Procedia engineering. 2017;**210**:120-125

[21] Wu D. Study on Temperature Rise of Hydration and Impermeability of Concrete Made of Demolished Concrete Lumps and Fresh Concrete. Guangzhou: South China University of Technology; 2018. (in Chinese)

[22] Liu Z. Experimental Investigation of Non-reinforced Concrete Pier Foundation. Shenyang, China: Northeastern University; 2009. (in Chinese)

[23] Liu Z, Zhu F, Shu Z. Research on horizontal capacity of widen-bottom non-reinforced concrete pier foundation on cohesive soil. Journal of Liaoning Technical University (Natural Science). 2008;**6**:850-852. (in Chinese)

[24] Jiang J, Gao G. Study on modifying coefficient of end-bearing resistance of large diameter belled piles. Rock and Soil Mechanics. 2006;**12**:2282-2288. (in Chinese)

[25] Dai Z. Gradation of coarse aggregate of dam concrete. Yangze River. 1994;**09**:14-20. (in Chinese)

[26] ASTM C39. Standard test method of compressive strength of cylindrical concrete specimens: American Society for Testing and Materials. 2001

[27] Benítez A, Bertone J, Civitillo P. Verification test research on flexural strength test method of concrete pavement brick. Building Block and Block Building. 2010;**05**:24-28. (in Chinese)

[28] SL352. Code of Test for Hydraulic Concrete. Beijing: China Building Industry Press; 2006

[29] Li J, Xiao J, Sun Z. Properties of recycled coarse aggregate and its influence on recycled concrete. Journal of Building Materials. 2004;**04**:390-395. (in Chinese)

[30] GB/T 50081. Standard test method for mechanical properties of ordinary concrete. Beijing: China Building Industry Press; 2002

Chapter 3

Vertical Bearing Capacity of Precast Pier Foundation Filled with Demolished Concrete Lumps

Bin Lei, Wengui Li, Zhuo Tang and Fuzhi Yang

Abstract

The application of recycled compound concrete made of demolished concrete lumps (DCLs) and fresh normal concrete in pier foundation can effectively improve the utilization efficiency of construction waste resources. In this study, two prefabricated pier foundations based on recycled compound concrete (dimension of Ø800 × 2500 mm and Ø1000 × 2500 mm) and two cast-in-place pier foundations based on ordinary concrete (dimension of Ø800 × 2500 mm and Ø1000 × 2500 mm) were tested. Special attention was devoted to the load-settlement curve characteristics of the precast pier foundation of compound concrete, the load transfer law of the pier-soil system, the soil pressure distribution at the bottom of the pier, and the failure mode. The results showed that the Q-S curve of precast concrete pier foundation made of recycled compound concrete is slow deformation at loading, which is consistent with that of cast-in-place concrete pier foundation. The load transfer theory of pier-soil system is established, and its accuracy is verified by experimental analysis. The precast foundation of recycled compound concrete is the same as the cast-in-place foundation of ordinary concrete. The failure form of prefabricated pier foundation made of recycled compound concrete was a local shear failure, while the failure form of ordinary concrete cast-in-place pier foundation was piercing-type shear failure. The feasibility of relevant theoretical methods for calculating the vertical ultimate bearing capacity is examined.

Keywords: recycled compound concrete, demolished concrete lump, pier foundation, vertical bearing capacity

1. Introduction

With the growing interest in sustainable and environmentally friendly construction, recycled aggregate concrete (RAC) has been widely investigated and is gradually being incorporated into real engineering projects [1–4]. A series of investigations has been conducted around the world on the mechanical properties, durability, and structural performance of RAC [2, 5–7]. Adopting demolished concrete blocks (DCBs) rather than recycled aggregates in structural members may reduce the cost of reuse of waste concrete. DCBs have distinctly larger size than the conventional recycled aggregates [8–11]. Therefore, the utilization of construction and demolition waste has great significance in saving resources, protecting the environment, and realizing the sustainable development of the construction industry. Combined with the favorable geological conditions of Nanchang city, this paper made a preliminary exploration on casting large-size demolished concrete lumps (DCLs) and fresh concrete, e.g., recycled compound concrete, into pier foundation, which was used as a new type of building foundation form.

The research on pier foundation was first carried out in the United States, followed by the former Soviet Union, Japan, and other countries. In particular, a large number of experimental results were obtained in the research on pier foundations with enlarged bottom [12]. The experimental research on pier foundation in China is relatively lagging behind. The theoretical research on pier foundation is more than the experimental research, mainly because it requires a large amount of financial and personnel resources to carry out the test. In terms of theoretical research, it mainly focuses on the soil failure theory [13], the design of pier foundation [14], the law of load transfer [15, 16], and the formula for calculating ultimate compressive bearing capacity [16–19]. As for experimental study, it includes the research on end resistance reduction [20], settlement deformation [21, 22], working mechanism [23–26], and bearing capacity formula [25, 27, 28]. Based on the experimental result and theoretical analysis, the paper provides a basis for the research on the vertical bearing capacity of the precast pier foundation with compound concrete.

In general, pier foundation is a kind of foundation form between rigid independent foundation and manual bored pile foundation [27, 29]. In multistory buildings, when the upper shallow foundation soil is relatively good and no weak layer beneath, when compared with shallow foundation, the artificial dig-hole pile foundation does not have the economic advantage, as it needs to use shallow foundation excavation of earthwork, which results in the high cost [27]. However, the pier foundation has good seismic performance, large single pier bearing capacity, and less earthwork excavation [28]; it can meet the design requirements of the building, and the foundation cost is low [29]. Hence, the research on the precast pier foundation made from recycled compound concrete has important engineering application value.

In this study, two prefabricated pier foundations based on recycled compound concrete (dimension of Ø800 × 2500 mm and Ø1000 × 2500 mm) and two cast-in-place pier foundations based on ordinary concrete (dimension of Ø800 × 2500 mm and Ø1000 × 2500 mm) were tested. Special attention was devoted to the load-settlement curve characteristics of the precast pier foundation of compound concrete, the load transfer law of the pier-soil system, the soil pressure distribution at the bottom of the pier, and the failure mode, and then the vertical bearing capacity of precast pier foundation filled with demolished concrete lumps is analyzed.

2. Geotechnical characteristics

The area is located in Jiangxi vocational technical college campus, Nanchang, Jiangxi province. The standard penetration test and the origin of a heavy dynamic penetration test were conducted to investigate the geotechnical characteristics. The soil samples collected by three apertures are shown in **Figure 1**.

According to the analysis of drilling soil samples in engineering geology, the soil stratigraphy consists of two main layers: Holocene filling soil (Q_4^{ml}) and Holocene alluvial layer (Q_4^{al}). According to its lithology and engineering characteristics, the distribution of each soil layer in the survey depth is described as follows:

Layer 1: silt (Q_3^{al}), brown and yellow, hard plastic, local hard, slightly dry, dry strength low, low toughness, shaking response medium, dull, components for clay,

Figure 1.
Soil samples collected from the test sites. (a) ZK_1, (b) ZK_2, and (c) ZK_3.

Soil layer name and number	Characteristic value of bearing capacity (kPa)	Soil heavy (KN/m^3)	Compression modulus E_s/deformation modulus E_0 (MPa)	Bored pile (clean bottom, D = 800 mm)	
				Ultimate lateral friction	Extreme resistance
				q_{sik} (kPa)	q_{pk} (kPa)
Silty soil	150	18	5.5/	50	—
Medium sand	160	18	/9.0	40	800

Table 1.
Suggested values of physical and mechanical parameters of bedrock soil layer.

sand. The layer thickness is 0.40–1.80 m, and the depth of the top layer is 0.00 m, distributed throughout the field

Layer 2: middle sand (Q_3^{al}), brown and yellow, loose to slightly dense, slightly wet, particle composition is mainly quartz, the upper part contains more shale, exposed layer thickness 5.40–5.90 m, layer top buried depth 0.40–1.80 m, distributed throughout.

The main soil properties of each soil layer, derived from the geotechnical investigation, and the evaluation of in situ and laboratory tests are presented in **Table 1**.

3. Experimental procedures

3.1 Raw materials

3.1.1 Concrete

Commercial concrete was used in this study as the fresh concrete, with the mix proportions shown in **Table 2**. The concrete used had a cement content of 220 kg/m^3, a water-cement ratio of 0.46, and a maximum aggregate grain size of 31.5 mm. The slump was 180 mm, and the compressive strength was 35.2 MPa.

Water-cement ratio	Water (kg/m^3)	Cement (kg/m^3)	Coarse aggregate (kg/m^3)	Fine aggregate (kg/m^3)	Fly ash (kg/m^3)	Water reducer (kg/m^3)
0.46	108	220	1050	872	120	6.3

Table 2.
Mix proportions of commercial concrete.

(a) (b) (c)

Figure 2.
Demolished concrete lumps. (a) Demolished concrete cubes, (b) crushing by an excavator, (c) concrete lumps selection.

3.1.2 Demolished concrete lumps

The DCLs used in the present study were obtained from the same batch of 150 mm cubic concrete samples reserved in the laboratory for more than 2 years. The cubic compressive strength of DCLs was 30.1 MPa. The demolished concrete cubes from an engineering quality inspection center were broken into lumps by excavator, as shown in **Figure 2**. Only concrete lumps with the sizes of 40–60 mm and 60–80 mm were selected for the inclusion into the recycled compound concrete, and these chosen lumps were used directly in the recycled compound concrete, i.e., without any further processing.

3.1.3 Reinforcement details

As shown in **Figure 3**, HRB 335 bar with a diameter of 12 mm was used for longitudinal bars of the pier foundation, while HRB 335 with a diameter of 8 mm was used for transverse reinforcement with a spacing of 250 mm. Four steel reinforcement cages were made, as shown in **Figure 3**, to serve as the structural reinforcement of the four pier foundations. Two of the piers are 800 mm in diameter and the other two are 1200 mm in diameter. All the four piers are 2400 mm in height. The longitudinal bar selected is HRB335 hot rolled thread steel bar with a diameter of 12 mm. The stirrup is the same type of steel reinforcement with a diameter of 8 mm, including 8 Φ 12 mm and longitudinal reinforcement stirrup Φ 8 @ 250 mm.

Figure 3.
Steel reinforcement cage.

Figure 4.
Molds for concrete casting.

3.1.4 Concrete molds

Concrete molds were prepared from wood, as shown in **Figure 4**, one with an inner diameter of 800 mm and a height of 2500 mm and the other with an inner diameter of 1000 mm and a height of 2500 mm.

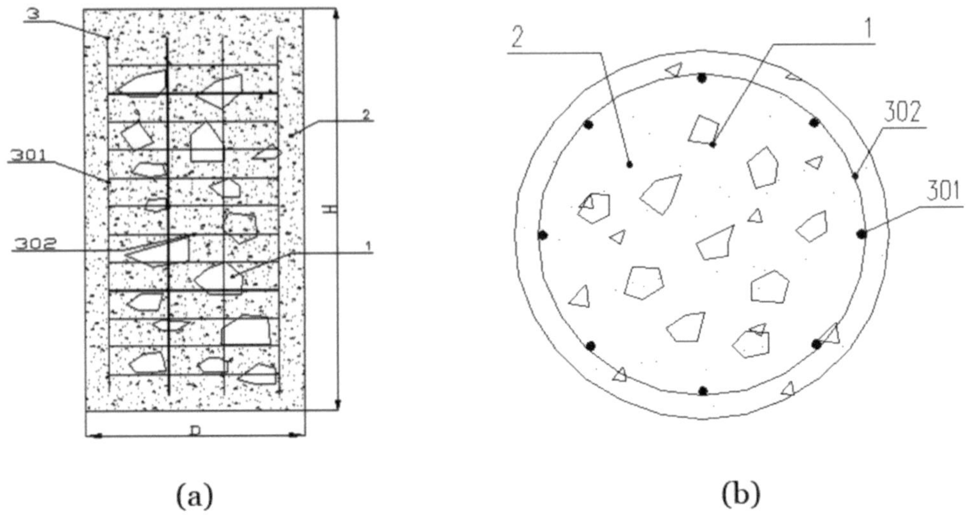

Figure 5.
Schematic diagram of precast pier foundation design (1, DCLs; 2, fresh concrete; 3, reinforcement; 301, longitudinal reinforcement; 302, transverse reinforcement). (a) Longitudinal section, (b) cross-section.

Notation	Count	Dimension (mm × mm)	DCLs grading	$f_{cu,new}$ (MPa)	$f_{cu,old}$ (MPa)	η (%)
YD800	1	Ø800 × 2500	40–60 mm:60–80 mm = 6:4	35.24	C25	30
YD1000	1	Ø1000 × 2500				
XD800	1	Ø800 × 2500	—	35.24	—	—
XD1000	1	Ø1000 × 2500	—			

Note: YD800 refers to the precast pier foundation of compound concrete with a diameter of 800 mm. XD1000 refers to the cast-in-place concrete pier foundation with a diameter of 1000 mm.

Table 3.
Details of pier foundations.

3.2 Pier foundation preparation

A total of four pier foundations were prepared, including two precast pier foundations filled with DCLs and fresh concrete (pier foundation dimension: Ø800 × 2500 mm, Ø1000 × 2500 mm) and two ordinary concrete cast-in-situ pier foundation (pier foundation dimension: Ø800 × 2500 mm, Ø1000 × 2500 mm. The schematic diagram of precast pier foundation design is shown in **Figure 5**. The details of pier foundations are presented in **Table 3**.

For pier foundations made of recycled compound concrete (YD1000 and YD800), the replacement rate of DCLs was 30%, and the ratio of DCLs with diameters of 40–60 mm to 60–80 mm was 6:4. The precast pier foundation was cast near the test site. After 28 days of curing, the pier foundations were transported to the construction site. In order to ensure the lifting of the pier foundations during transportation, it is necessary to embed the lifting rings in the pier foundations before they are cast. As suggested in the literature [30], the bending center diameter of the lifting ring was set as 63 mm, the burial depth was set at 75 mm, and the bending hook was set as 25 mm. The casting process of the pier foundation is as follows:

1. Prepare reinforcement cages with diameters of 0.7 and 0.9 m for the pier foundations with diameters of 0.8 and 1.0 m, respectively. Strain gauges were pasted on the longitudinal reinforcement, and the layout of the strain gauges are presented in **Figure 6**. Moreover, the strain gauges were pasted to the outer sides of the longitudinal bars to prevent damage of the strain gauge during concrete casting and vibration.

2. Assemble and splice the wood mold with an inner diameter of 0.8, 1 m, and length of 2.5 m, and drill a hole with diameter 50 mm at the upper part of the mold as the outlet of strain gauge cables.

3. When the height of the mold was 2.5, scaffolding was set up for construction in the casting project. On the one hand, it could fix the mold; on the other hand, it could serve as an operating platform during the concrete casting process.

4. Put the steel reinforcement cage into the mold, and adjust the spacing between the steel reinforcement cage and the mold that there was a 50 mm gap between. The DCLs were wetted before casting to prevent the DCLs from absorbing too much water in the casting process. The fresh concrete is first casted to the bottom of the mold with a height of 250 mm as a cushion. The fresh concrete incorporation with the DCLs with the replacement rate of 30% was casted into the mold. During the casting process, the handheld vibrator is used for continuous vibration. When the casted layer reached 500 mm high, a steel wire mesh with a diameter slightly smaller than the diameter of the steel wire cage was laid on the fresh concrete. Repeat and alternate the above casting process until the mold is filled. The schematic diagram for the height setting of layered steel mesh is shown in **Figure 7**. The steps for the

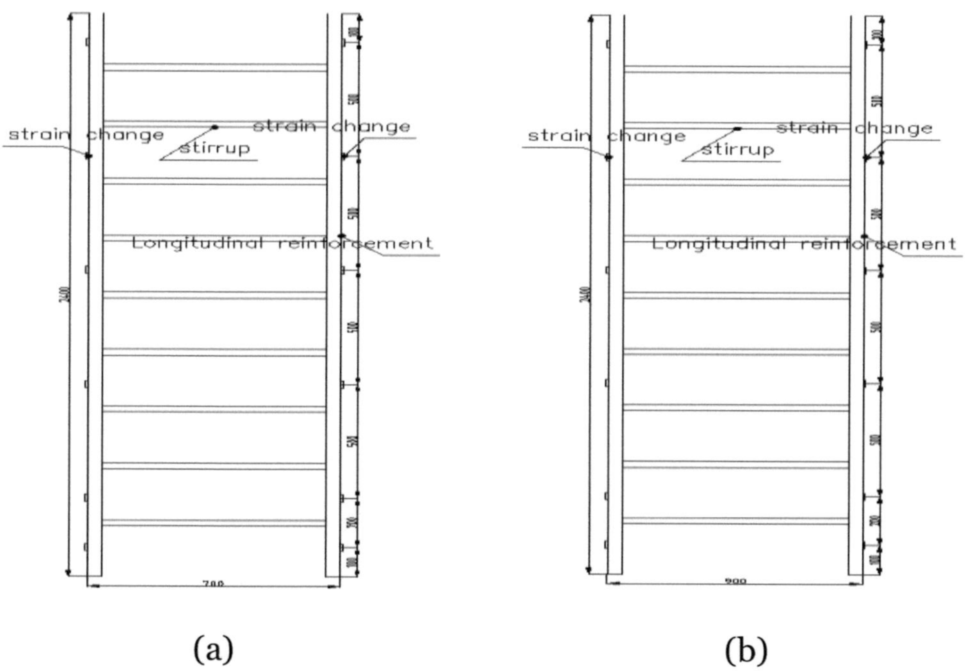

Figure 6.
Layout of strain gauge. (a) Pier foundations of YD800 and XD800, (b) pier foundations YD1000 and XD1000.

preparation of recycled compound concrete and ordinary concrete piers are shown in **Figures 8** and **9**, respectively. After 1 day of curing, the concrete was demolded. Carpet saturated with water was used to wrap the surface of the pier foundations. After 28 days of curing, the pier foundation was lowered into the hole, a little larger than the pile diameter.

5. The gap between the precast pier foundation and the hole shall be backfilled and compacted with undisturbed soil. At the same time, the hanging ring of the precast pier top was sawed off with the cutting machine to ensure the levelness of the pier top.

6. As the two cast-in-place piers of ordinary concrete are casted on the foundation, the pier foundation preparation is as follows. The test hole was dug firstly. Then, the steel cage was lowered into the hole, which was followed by fresh concrete casting. After casting, cover the top of the pier foundation with plastic film to prevent water evaporation.

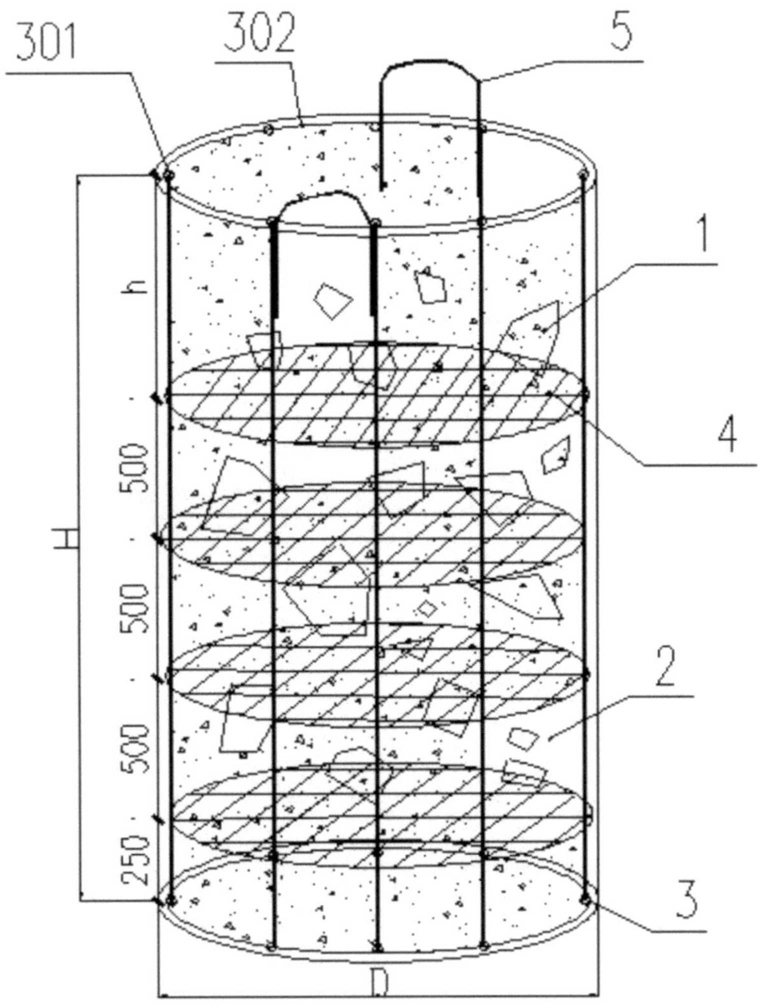

Figure 7.
Schematic diagram of the casting process (1, DCLs; 2, ordinary concrete; 3, reinforcement; 301, longitudinal reinforcement; 302, transverse reinforcement; 4, steel wire mesh; 5, rings).

Figure 8.
Construction steps of precast pier foundation filled with DCLs and fresh concrete for foundation casting. (a) Prepare reinforcement, (b) install mold, (c) set up scaffolding and molds, (d) put the steel cage into the mold, (e) cast new concrete at the bottom of the mold and then place a steel wire mesh, (f) cast new concrete incorporated with concrete lumps to the predetermined height, (g) consolidate concrete, (h) place the second layer of wire mesh, (i) cast new concrete with concrete lumps to the predetermined height and then consolidate, (j) follow steps h and i until the mold is filled, (k) demold after 1 day of curing, (l) cure pier foundation, (m) lower pier foundation into the hole, (n) fill the gap between prefabricated pier foundation and hole with undisturbed soil, (o) saw off the lifting ring to ensure the levelness of pier top end.

Figure 9.
Construction steps of cast-in-place ordinary concrete piers for foundation casting. (a) Prepare reinforcement; (b) manually dig the hole; (c) put the steel cage into the hole, followed by concrete casting and consolidation; (d) seal the top with membrane to prevent evaporation of water.

3.3 Testing setup

The commonly used reaction methods for vertical compression and static load test of pier foundation include the platform reaction beam method, anchor pile reaction beam method, and anchor pile reaction beam loading method [31]. Based on the actual situation of the site, the platform reaction beam method was adopted in this study, as shown in **Figure 10**. Load counterforce device consisted of the bearing plate, reaction beam system, 320 tons of double oil jack, and 63 MPa ultrahigh pressure pump station. Load counterforce device could provide the reaction force not less than 1.2 times of maximum load value. The load was applied at a rate of movement corresponding to a stress rate on the pier foundation as specified in the code [32]. The compression force was measured by a static load tester (type: RS-JYB), along with the settlement measured by two displacement sensors. To

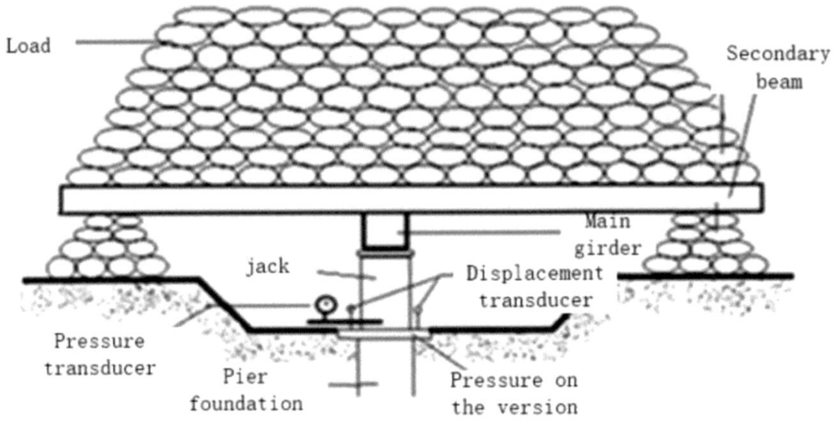

Figure 10.
Test setup.

Model	Dimension	Scale range	Resolution (% F.S.)	Temperature measurement range (°C)	Temperature measurement precision (°C)
JTMV2000	Ø117 × 28 mm	0–2.5 MPa	≤ 0.08	−25 to 60	0.5

Note: F.S. refers to the full range of the equipment.

Table 4.
Parameters of earth pressure gauges.

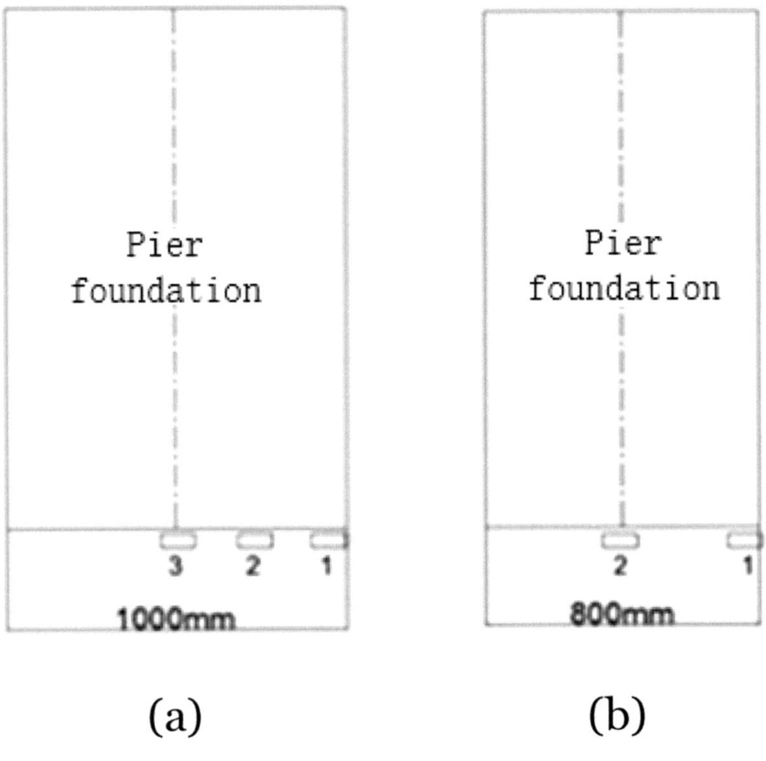

Figure 11.
Arrangement of earth pressure boxes. (a) Pier foundation with a diameter of 1000 mm, (b) pier foundation with a diameter of 800 mm.

understand the stress distribution along the pier foundation, six pairs of strain gauges were arranged along with the longitudinal reinforcement, as shown in **Figure 6**. To understand the soil pressure distribution of the foundation, the vibrating wire pressure sensors (JTM-v2000) were applied. The parameters of the sensor are shown in **Table 4**. The vibrating wire pressure sensors were arranged at 100 mm intervals at the bottom of the pier foundation, and the distribution of the earth pressure at the bottom of the pier foundation can be measured by considering half of the symmetrical arrangement. Two earth pressure boxes are required for the diameter of 0.8 m, and three earth pressure boxes are required for the diameter of 1 m. The detailed installation of vibrating wire pressure sensor is shown in **Figure 11**.

4. Results and analysis

4.1 Load-settlement relationship

The measured load-settlement relationships are plotted in **Figure 12**, and the characteristic values are listed in **Table 5**. All the pier foundations exhibit similar load-settlement curves, which belongs to slow variation. According to the geotechnical characteristic measured, the bearing layer of the pier foundation test was located in the middle sand layer, which is consistent with the Q-S curve characteristics of sandy soil [33]. Regardless of the type of pier foundation (recycled compound concrete or ordinary concrete), the characteristic value of vertical bearing capacity of single pier increases with the increase of the diameter of pier foundations. For pier foundations with the same diameter, the vertical bearing capacity of the recycled concrete precast pier is lower than that of the ordinary concrete pier by 25–35%. This is mainly because ordinary concrete was cast-in-situ, and therefore there was relatively large friction between the pier foundation and the soil, while

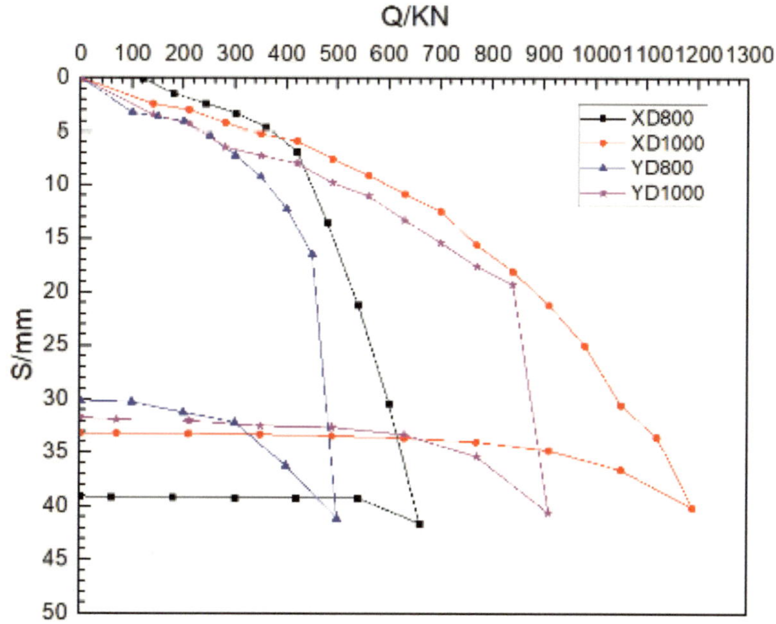

Figure 12.
Q-S curve of the foundation of each pier.

Pier foundation	Q_{cr} (kN)	S_{max} (mm)	Q_u (kN)	S_{qu} (mm)	R_a (kN)	S_{ra} (mm)	Rebound deformation (%)	$\Delta s_e/\Delta s_{e-1}$
YD800	500	41.15	450	16.42	225	4.02	26.8	5.9
XD800	660	41.55	660	41.55	330	3.22	6.0	1.2
YD1000	910	40.56	840	19.25	420	7.95	21.8	12.6
XD1000	1190	40.13	1120	33.50	560	9.09	17.4	2.21

Note: Q_{cr} represents the maximum loading value of pile-soil system failure; S_{max} represents the total settlement at the maximum loading; Q_u represents the ultimate vertical compression bearing capacity of a single pier; S_{qu} means the settlement corresponding to the ultimate bearing capacity; R_a represents the characteristic value of vertical bearing capacity of single pier; S_{ra} represents the settlement corresponding to the characteristic value of bearing capacity; $\Delta s_e/\Delta s_{e-1}$ represents the ratio of the absolute settlement generated by the last loading level to that generated by the upper loading level.

Table 5.
Comparison of foundation analysis data of piers.

for precast pier foundation made of recycled compound concrete, the friction generated between the pier foundation surface and the soil can be ignored. Moreover, the rate of rebound deformation of the precast recycled compound concrete is higher than that of the cast-in-place ordinary concrete pier after unloading. It is mainly due to the precast pier, where the friction between the pier and the soil was insignificant, and consequently, the load transformed to the pier end was relatively larger. Upon the maximum loading, the precast recycled compound concrete pier had larger settlement than the cast-in-situ ordinary concrete pier. As shown in **Table 5**, the $\Delta s_e/\Delta s_{e-1}$ of YD1000 is 12.6, while that of XD1000 is 2.21, indicating that the foundation failure mode of the precast compound concrete pier is different from that of a cast-in-place ordinary concrete pier.

4.2 Load transfer law of pier-soil system

4.2.1 Geological distribution of the test site

The geological distribution of the pier foundation test in the test site is silt in the range of 0.00–0.40 m and sandy soil in the range of 0.40–5.90, respectively.

4.2.2 Load transfer theory of pier-soil system

Based on the load transfer principle of large-diameter single pile without expanding bottom, and relevant literature [34, 35], the load transfer between the pier-soil system includes the following three stages:

1. At the initial stage, the load is mainly undertaken by the friction resistance between pier foundation and soil. The friction resistance of soil at the bottom of the pier does not occur, because of the relative displacement between the bottom of the pier and soil.

2. With the increase of the load, the compression deformation of the whole pier body and the relative displacement between the piers will increase, and thus the friction resistance at the bottom of the pier foundation will gradually increase, and the soil layer at the bottom of the pier will also experience compression settlement with the increase of pier end stress, and the pier end provides part of end resistance.

3. When the load is further increased, the settlement of the pier bottom increases, and correspondingly, the end resistance provided by the pier end increases significantly. On the other hand, with the compression of the soil layer at the pier end, the relative displacement of the piers also increases, and the friction resistance provided by the pier side will further increase.

4.2.3 Analysis of axial force and average friction resistance of pier body

Based on the mechanics of materials that the material satisfies Hooke's law in the elastic deformation range, the relationship between stress and strain could be expressed as Eq. (1).

$$\sigma_c = E_c \times \varepsilon_c \tag{1}$$

where σ_c is the stress of concrete pier foundation, ε_c is the strain of concrete pier foundation, and E_c is the elastic modulus of concrete pier foundation. And the load can be calculated based on the following relationship.

$$N = \sigma_c \times A \tag{2}$$

If the linear expansion coefficient of steel and concrete could be determined by the same deformation together, the cross-section of concrete and steel bar would have the same strain value [36]. In this work, strain gauges placed on the longitudinal bars could serve the internal structure of the pier foundation. Along with the structural reinforcement, six sections are selected on both sides, and six strain gauges are placed on each side (**Figure 6**). By measuring the strain value, the axial force at the section is calculated, and then the distribution of friction resistance at the pier side can be obtained. Under the different levels of load, the axial force distribution diagram of pier body is presented in **Figure 13**.

As shown in **Figure 13**, it can be seen that with the increase of loading load, the axial force of the pier foundation increases. The axial force of the cast-in-place concrete pier foundation decreases with the depth of the pier foundation, indicating that the friction resistance at the pier side plays an important role. When the pier-soil system is damaged, the side resistance of the pier accounts for around 25% of the ultimate compressive bearing capacity. For the precast pier, the lateral resistance is almost zero, which is mainly caused by the gap between the precast pier foundation of compound concrete and the surrounding holes are not easy to be compacted when the undisturbed soil is backfilled.

The friction resistance of the pier body of the cast-in-situ pier foundation is analyzed in the following way: as the height of the test pier foundation is 2.5 m, and the soil layer distributed around the pier foundation has been presented that 0-0.4 m of the pier body is slit, and 0.4–2.5 m of the pier body is medium sand. Under the maximum load, the average friction resistance of the pier body in different soil layers is distributed along the pier body, as shown in **Figure 14**. In addition, with the increase of pier diameter, the greater is the friction resistance of the pier foundation at the same depth, the greater is the vertical compression bearing capacity of the pier foundation.

4.3 Analysis of soil pressure distribution of pier bottom

According to the measurement of the soil pressure distribution of pier bottom foundation, vibrating string earth pressure gauge. During the test, data were read for three to four times with vibrating string sensor readout instrument when each

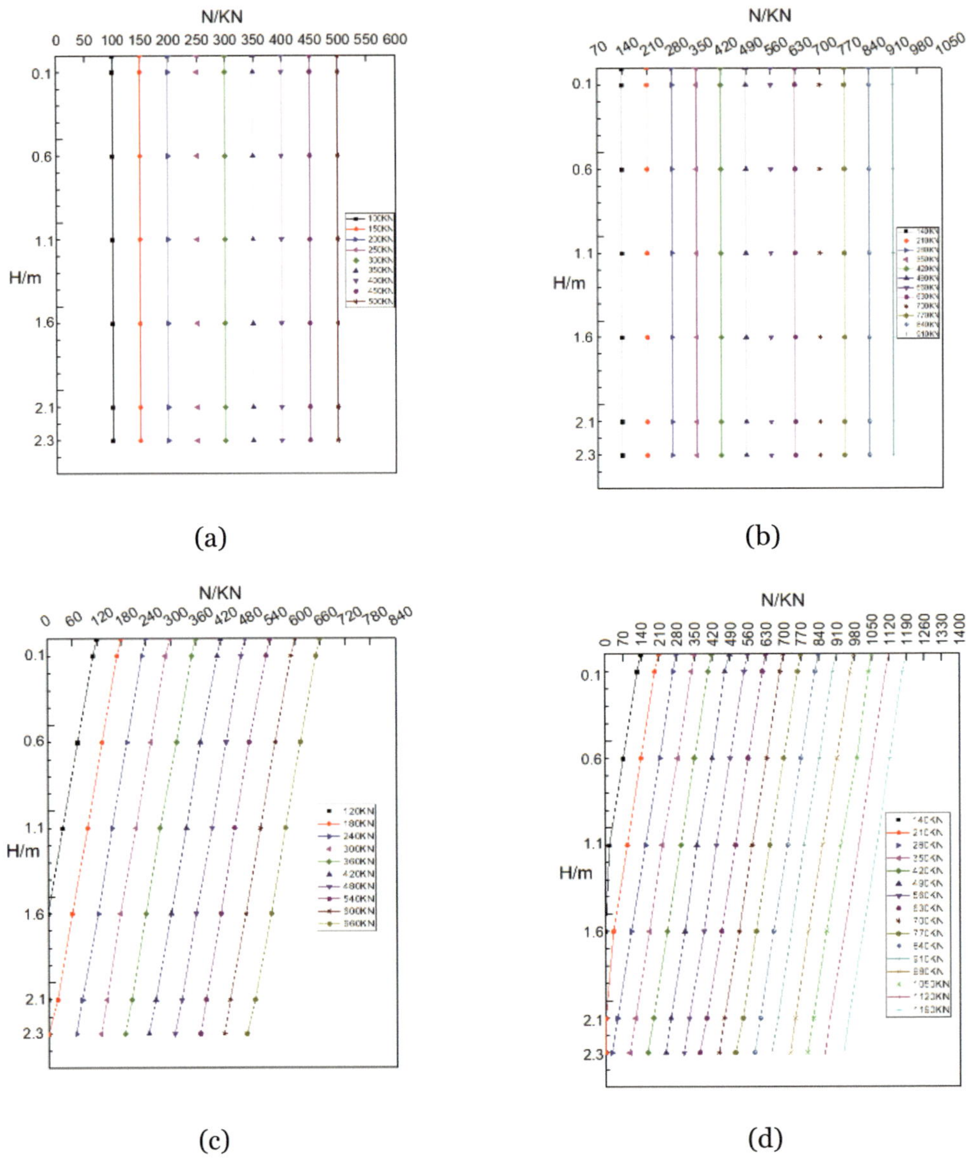

Figure 13.
Axial force distribution of piers. (a) YD800, (b) YD1000, (c) XD800, (d) XD1000.

level of the load was reached, and its average value was taken. The recorded frequency was converted into pressure through Eq. (3):

$$P = K\left(f_i^2 - f_0^2\right) \tag{3}$$

where P is the earth pressure value (MPa), K is the calibration coefficient of the earth pressure gauge (MPa/HZ^2), f_i is the frequency recorded during loading (HZ), and f_0 is the initial frequency (HZ).

Figure 15 presents the measured earth pressure distribution of the foundation at the bottom of the pier. The distribution of soil pressure of precast pier foundation of compound concrete and that of cast-in-place pier foundation of ordinary concrete has the same pattern, that is, the soil pressure of foundation of pier bottom is distributed in a parabolic shape. The main reason is that the bearing layer of the pier

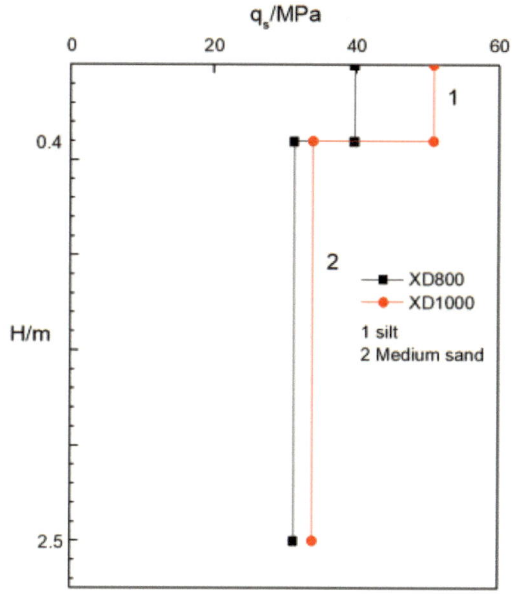

Figure 14.
Average frictional resistance distribution of cast-in-situ concrete piers.

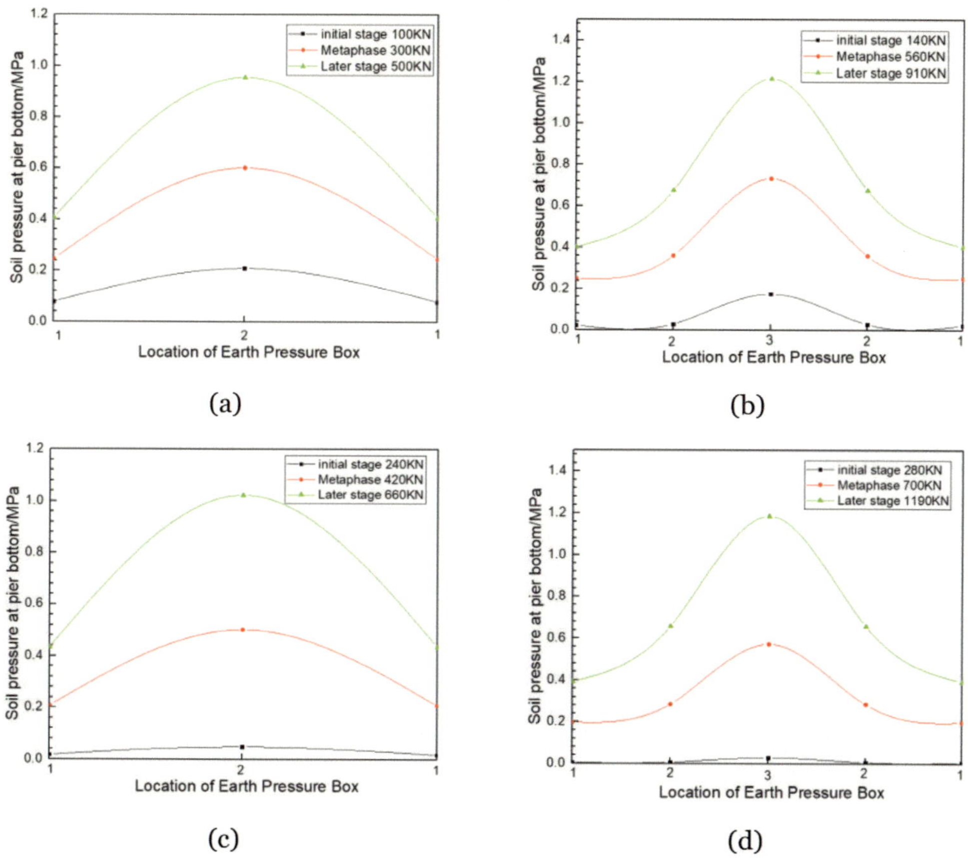

Figure 15.
Distribution of foundation soil pressure at the beginning, middle, and later stage of loading. (a) YD800, (b) YD1000, (c) XD800, (d) XD1000.

foundation was on the sand soil. In the process of loading, the sand grains at the bottom edge of the pier are easily extruded laterally, and the soil pressure at the edge of the foundation transfers to the middle part, leading to the stress redistribution. As a result, the soil pressure at the edges of the pier bottom was small, while the soil pressure at the middle of the pier bottom was large. At the initial stage of loading, the soil pressure value of the prefabricated pier foundation is much higher than that of the cast-in-place compound concrete pier foundation with the same diameter, which further indicates that the prefabricated pier foundation with recycled compound concrete is almost entirely provided by the resistance at the pier end. Also, with the increase of load, the failure of the bottom edge of the pier was earlier than that of the middle of the pier bottom.

4.4 Failure mode analysis of the pier foundation

The failure modes are presented in **Figure 16**, which the soil of both precast pier foundation and ordinary concrete cast-in-situ pier did not exhibit the bulging damage phenomenon. For precast concrete pier foundation, the settlement increased obviously in the former period, and Q-S curve has obvious turning point (**Figure 12**); therefore, along with the failure mode, the failure belongs to local shear failure damage. For ordinary concrete pier, as the load increases, the soil at the bottom of the pier foundation was compressed. When loaded into the pier-soil system damage, the rate of settlement remained unchanged, and Q-S curve did not show obvious turning point; therefore, along with the failure mode, the failure belongs to piercing-type shear failure forms.

4.5 Research on vertical bearing capacity formula

For the prediction of vertical bearing capacity of pier foundation, according to the analysis of load transfer law of pier-soil system in Section 3.2, it can be seen that the stress form of unexpanded bottom pier foundation tested in this paper is theoretically consistent with that of large-diameter pile foundation. Therefore, the calculation formula of vertical bearing capacity can refer to the code for building pile foundation technology (JGJ 94-2008) [37], specification for design of building

(a) (b)

Figure 16.
Failure modes of concrete piers. (a) Soil around the pier when the precast pier foundation of recycled compound concrete was destroyed, (b) soil around the piers when the foundation of the cast-in-place ordinary concrete was destroyed.

foundation (GB 50007-2011) [38], and literature [29]. In the following section, the feasibility of these calculation formulas is investigated.

4.5.1 Calculation formula of technical code for building pile foundation (JGJ 94-2008)

The formula for calculating the ultimate bearing capacity of a large-diameter single pile can be calculated according to Eq. (4):

$$Q_{uk} = Q_{sk} + Q_{pk} = u\sum \psi_{si} q_{sik} l_i + \psi_p q_{pk} A_p \tag{4}$$

where Q_{sk} stands for total ultimate resistance measured at pile side (kN), Q_{pk} represents the total limit end resistance (kN) of pile side, u is the perimeter (m) of pile body, Ψ_{si} represents the dimensional reduction coefficient of pile lateral resistance (refer to **Table 6**), q_{sik} represents the ultimate lateral resistance (kPa) of the pile side in layer I soil, Ψ_p represents the reduction coefficient of pile tip resistance (refer to **Table 6**), and q_{pk} represents the ultimate end resistance (KPa) with a pile diameter of 800 mm. The reduction coefficients Ψ_{si} and Ψ_p in the formula are calculated according to **Table 6.**

4.5.2 Calculation formula of code for design of building foundation (GB 50007-2011)

The preliminary design of the vertical bearing capacity characteristic value of a single pile is estimated according to Eq. (5):

$$R_a = q_{pa} A_p + u_p \sum q_{sia} l_i \tag{5}$$

where q_{pa} represents the characteristic value of pile tip resistance (kPa), A_p represents the area of pile bottom (m^2), q_{sia} represents the characteristic value of pile lateral resistance (kPa), u_p represents the pile girth (m), and l_i represents the thickness of soil layer i (m).

However, the vertical ultimate bearing capacity of a single pile is divided by safety factor 2, that is, the characteristic value R_a of vertical bearing capacity of a single pile. Therefore, based on this, $2R_a$ is adopted to calculate the vertical ultimate bearing capacity of a single pile.

4.5.3 Calculation formula in literature

Based on the study in [29], the characteristic value of the vertical bearing capacity of single pier can be calculated by referring to Eq. (6):

$$R_a = \beta f_a A_D \tag{6}$$

where β is the pier foundation correction coefficient, f_a represents the modified characteristic value of foundation bearing capacity of pier base (KPa), and A_D

Soil type	Sand, gravel soil
Ψ_{si}	$(0.8/d)^{1/3}$
Ψ_p	$(0.8/D)^{1/3}$
Note: equal diameter pile d = D.	

Table 6.
Dimension reduction factor.

represents the cross-section area of pier, (m2). For the coefficient β, it could be referred to the values in **Table 7**.

The modified characteristic value of foundation bearing capacity of pier base (f_a) in Eq. (6) could be calculated according to Eq. (7):

$$f_a = f_{ak} + \eta_d \cdot \gamma_0 (d - 1.5) \tag{7}$$

where f_{ak} represents the characteristic value of foundation bearing capacity (kPa), η_d refers to the correction coefficient of bearing capacity of foundation considering the buried depth of pier foundation, and γ_0 denotes the depth of the bottom soil of the pier foundation. For groundwater, its effective depth (kN/m^2) d represents the embedded depth of pier foundation (m).

Additionally, the vertical ultimate bearing capacity of a single pile is divided by the safety factor 2, that is, the characteristic value R_a of the vertical ultimate bearing capacity of a single pile. Therefore, $2R_a$ is used to calculate the vertical ultimate bearing capacity of a single pile.

4.5.4 Pier foundation calculation formula selection

Based on Eqs. (4)–(6), **Table 8** shows the comparison of the predicted values and tested results. For the value of Q_{us}/Q_{uj} greater than 1, it indicates that the test value is greater than the calculated value, while for the value of Q_{us}/Q_{uj} less than 1, it indicates that the vertical bearing capacity of pier foundation will be utilized fully

H/D	1.6	1.8	2.0	2.2	2.4	2.6	2.8	3.0
Sandy soil, gravel soil	1.6	1.7	1.8	1.9	2.1	2.3	2.5	2.7

Note: H represents the height of pier foundation, m; D is the diameter of pier foundation, m.

Table 7.
Pier foundation correction coefficient.

Formula	YD800			YD1000			XD800			XD1000		
	Q_{uj}	Q_{us}	Q_{us}/Q_{uj}	Q_{uj}	Q_{us}	Q_{us}/Q_{uj}	Q_{uj}	Q_{us}	Q_{us}/Q_{uj}	Q_{uj}	Q_{us}	Q_{us}/Q_{uj}
JGJ94-2008	400	450	Q1.125	582	840	1.443	661	660	0.998	885	1120	1.265
GB50007-2011	400	450	1.125	628	840	1.337	661	660	0.998	954	1120	1.174
Literature [64]	432	450	1.041	678	840	1.238	645	660	1.023	1013	1120	1.105

Note: Q_{uj} represents the value calculated by the reference formula of vertical ultimate bearing capacity, kN; Q_{us} represents the test value of vertical ultimate bearing capacity under compression, kN.

Table 8.
Comparison of predicted values and test values.

Formula	YD800	YD1000	XD800	XD1000	Average	Mean square error	Coefficient of variation
JGJ94-2008	1.125	1.443	0.998	1.265	1.207	0.191	0.158
GB50007-2011	1.125	1.337	0.998	1.174	1.158	0.140	0.121
Literature [18]	1.041	1.238	1.023	1.105	1.101	0.09	0.09

Table 9.
Statistical characteristics of Q_{us}/Q_{uj} ratio under different calculation formulas.

in the design process. However, it should be noted that the minimum bearing capacity cannot be less than 0.5. Otherwise there will be safety risks and accidents might occur.

The statistical characteristic of the ratio between the test value and the calculated value by different methods are shown in **Table 9**. It can be seen from **Table 9** that the variation coefficient of the formula by Liu [29] is the minimum, indicating that the calculated value of the formula has the minimum fluctuation compared with the experimental value. At the same time, its average value is also the smallest, which is greater than 1, indicating that it is appropriate to calculate the safety reserve.

5. Conclusion

Based on the investigations on the vertical bearing capacity of precast pier foundation filled with demolished concrete lumps, the following conclusions can be drawn:

1. The Q-S curve of prefabricated pier foundation made of recycled compound concrete is of slow deformation when loaded, which is consistent with that of cast-in-place concrete pier foundation. Through Q-S curve, the vertical bearing capacity of prefabricated pier foundation with recycled compound concrete is about 25–30% lower than normal concrete cast-in-situ pier foundation, and in the pier foundation of the same diameter, the rebound deformation of prefabricated pier foundation was higher than that of cast-in-situ pier.

2. With the increase in diameter, the characteristic value of the vertical bearing capacity of the prefabricated foundation made of recycled compound concrete increases. At the same time, when the vertical ultimate bearing capacity of the single pier is reached, the settlement value also increases, which is consistent with the performance law of the cast-in-situ ordinary concrete pier.

3. The load transfer theory of pier-soil system was established, and the accuracy of the theory was verified through experimental analysis. At the same time, the variation rules of axial force, average friction resistance, and the similarities and differences between cast-in-place concrete pier foundation and prefabricated pier foundation made of recycled compound concrete are analyzed by placing strain gauge on longitudinal bars of pier body structure.

4. By placing the earth pressure box at the bottom of the pier, it is analyzed that the prefabricated foundation of DCLs is the same as the cast-in-place foundation of the ordinary concrete. The earth pressure at the bottom of the pier is distributed in a parabola.

5. Based on the comprehensive analysis of the soil pressure distribution and Q-S curve change rule of the foundation at the bottom of the pier, the failure form of the soil mass of the prefabricated pier foundation of recycled compound concrete is local shear failure, while the failure form of the soil mass of the cast-in-situ pier foundation of ordinary concrete is piercing-type shear failure.

6. The method proposed in this study provides a reasonable prediction for the vertical compressive ultimate bearing capacity for both prefabricated pier foundation with recycled compound concrete and cast-in-situ pier foundation of ordinary concrete.

Acknowledgements

This work was supported by the National Natural Science Foundation of China (51668045 and 51808133) and Jiangxi Science and Technology Committee (20161BBG70056). The authors also acknowledge the research support from the Australian Research Council (DE150101751), Australia.

Author details

Bin Lei[1,2], Wengui Li[2*], Zhuo Tang[2] and Fuzhi Yang[2]

1 School of Civil Engineering and Architecture, Nanchang University, Nanchang, Jiangxi, P.R. China

2 Center for Green Technology, School of Civil and Environmental Engineering, University of Technology Sydney, NSW, Australia

*Address all correspondence to: wengui.li@uts.edu.au

References

[1] Xiao J, Li W, Fan Y, Huang X. An overview of study on recycled aggregate concrete in China (1996-2011). Construction and Building Materials. 2012;**31**:364-383. DOI: 10.1016/j.conbuildmat.2011.12.074

[2] Li W, Sun Z, Luo Z, Shah S. Influence of relative mechanical strengths between new and old cement mortars on the crack propagation of recycled aggregate concrete. Journal of Advanced Concrete Technology. 2017;**15**(3):110-125. DOI: 10.3151/jact.15.110

[3] Lei B, Li W, Tam V, Sun Z. Investigation on properties of RAC under coupling loading and freeze-thaw cycles in salt-solution. Construction and Building Materials. 2018;**163**:840-849. DOI: 10.1016/j.conbuildmat.2017.12.194

[4] Li W, Luo Z, Long C, Wu C, Duan W, Shah S. Effects of nanoparticle on the dynamic behaviors of RAC under impact loading. Materials and Design. 2016;**112**:58-66. DOI: 10.1016/j.matdes.2016.09.045

[5] Lei B, Li W, Li Z, Wang G, Sun Z. Effect of cyclic loading deterioration on concrete durability: Freeze-thaw and carbonation. Journal of Materials in Civil Engineering. 2018;**30**(9):04018220. Available from: http://hdl.handle.net/10453/131532

[6] Li W, Xiao J, Sun Z, Kawashima S, Shah S. Interfacial transition zones in recycled aggregate concrete with different mixing approaches. Cement and Building Materials. 2012;**35**:1045-1055. DOI: 10.1016/j.conbuildmat.2012.06.022

[7] Li W, Chu L, Tam W, Poon C, Duan W. Effects of nano-particles on failure process and microstructural properties of recycled aggregate concrete. Construction and Building Materials. 2017;**142**:42-50. DOI: 10.1016/j.conbuildmat.2017.03.051

[8] Wu B, Jian S, Zhao X. Structural behavior of steel-concrete partially encased composite columns containing demolished concrete lumps under axial compression. Engineering Structures. 2019;**197**(15):1-18. DOI: 10.1016/j.engstruct.2019.109383

[9] Wu B, Yu Y, Zhao X. Residual mechanical properties of compound concrete containing demolished concrete lumps after exposure to high temperatures. Fire Safety Journal. 2019;**105**:62-78. DOI: 10.1016/j.firesaf.2019.02.008

[10] Wu B, Yu Y, Chen Z, Zhao X. Shape effect on compressive mechanical properties of compound concrete containing demolished concrete lumps. Construction and Building Materials. 2018;**187**:50-64. DOI: 10.1016/j.conbuildmat.2018.07.086

[11] Wu B, Li Z. Mechanical properties of compound concrete containing demolished concrete lumps after freeze-thaw cycles. Construction and Building Materials. 2017;**155**:187-199. DOI: 10.1016/j.conbuildmat.2017.07.150

[12] Zhou L. The study about of belled pier found on deformation modulus of bearing stratum [thesis]. Xi 'an University of Architecture and Technology; 2013 (in Chinese)

[13] Zhao Z, Fu Z. Discussion of working mechanism on pier foundation. Soils and Foundations. 2004;**2**:49-50 (in Chinese)

[14] Zhang J, Wang J. Analysis is and design of belled pier foundation. Chinese Journal of Geotechnical Engineering. 1996;**3**:67-73 (in Chinese)

[15] Wang F, Wang M. Working mechanism and bearing capacity design

of the pedestal pier with large diameter. Chinese Journal of Geotechnical Engineering. 2002;**2**:251-253 (in Chinese)

[16] Hoang N, Liao K, Tran X. Estimation of scour depth at bridges with complex pier foundations using support vector regression integrated with feature selection. Journal of Civil Structural Health Monitoring. 2018; **8**(3):431-442. DOI: 10.1007/s13349-018-0287-2

[17] Liu Y, Guo L, Zhang Q, et al. Study on the calculation method of vertical bearing capacity of pier foundation. Sichuan Building Materials. 2010;**36**(4): 50-52 (in Chinese)

[18] Ha J, Ko K, Jo S, Park H, Kim D. Investigation of seismic performances of unconnected pile foundations using dynamic centrifuge tests. Bulletin of Earthquake Engineering. 2019;**17**(5): 2433-2458. DOI: 10.1007/s10518-018-00530-y

[19] Sego D, Biggar K, Wong G. Enlarged base (belled) piles for use in ice or ice-rich permafrost. Journal of Cold Regions Engineering. 2003;**17**(2):68-88. DOI: 10.1061/(ASCE)0887-381X(2003)17:2(68)

[20] Jiang J, Gao G. Study on modifying coefficient of end-bearing resistance of large diameter belled piles. Rock and Soil Mechanics. 2006;**12**:2282-2288 (in Chinese)

[21] Liu Z, Jiang Y, Gong W. Discussion of the design method of large diameter belled pile's bearing capacity according to settlement. Journal of Southeast University (Natural Science Edition). 2001;**4**:49-53 (in Chinese)

[22] Lee C. Settlement and load distribution analysis of underreamed piles. ARPN Journal of Engineering and Applied Sciences. 2007;**2**(4):35-40

[23] Liu Z, Zhu F, Lu T, et al. Bearing capacity test of pier foundation in cohesive soil. Journal of Shenyang Jianzhu University Natural Science. 2008;**5**:783-787 (in Chinese)

[24] Liu Z, Shan M, Wang S, et al. Research of pier foundation on cohesive soil. Concrete. 2009;**2**:42-44 (in Chinese)

[25] Morici M, Minnucci L, Carbonari S, Dezi F, Leoni G. Simple formulas for estimating a lumped parameter model to reproduce impedances of end-bearing pile foundations. Soil Dynamics and Earthquake Engineering. 2019;**121**: 341-355. DOI: 10.1016/j.soildyn.2019.02.021

[26] Peter J, Lakshmanan N, Devadas MP. Investigations on the static behavior of self-compacting concrete under-reamed piles. Journal of Materials in Civil Engineering. 2006;**18**(3): 408-414. Available from: https://ascelibrary.org/doi/10.1061/(ASCE)0899-1561(2006)18:3(408)

[27] Liu Z, Zhu F, Shu Z, et al. Research on horizontal bearing capacity of widen-bottom non-reinforced concrete pier foundation on cohesive soil. Journal of Liaoning Technical University(Natural Science Edition). 2008;**6**:850-852 (in Chinese)

[28] Liu Z. Experimental study on bearing capacity of pier-type foundation with sand soil and gravel. Geotechnical Investigation & Surveying. 1996;**3**:9-12 (in Chinese)

[29] Liu Z. Experimental investigation of non-reinforced concrete pier foundation [thesis]. Nanjing: Northeastern University; 2009 (in Chinese)

[30] JGJ225T-2010. Technical specification for large diameter bored pile. Beijing: China Building Industry Press; 2010 (in Chinese)

[31] Zhang J, Wu Y, Xu Y, et al. Basic engineering testing technology and case analysis. Beijing: Science Press; 2012 (in Chinese)

[32] JGJ106–2014. Technical specification for foundation pile inspection[S]. Beijing: China Building Industry Press; 2014 (in Chinese)

[33] Wang Z. Comparison of design between manual bored pile and pier foundation. Sichuan Architecture. 2010; **30**(06):94-95 (in Chinese)

[34] Zhou H, Chen Z. Analysis of effect of different construction methods of piles on the end effect on skin friction of piles. Frontiers of Architecture and Civil Engineering in China. 2007;**1**(4): 458-463

[35] Huo Z. Calculation of vertical bearing capacity of pile with manual hole digging and bottom expanding [thesis]. Xi an: Chang 'an university; 2002 (in Chinese)

[36] Tong J, Lin S, Dai Y. Load transfer behavior of large bore bored piles. Chinese Journal of Geotechnical Engineering. 1994;**06**:123-131 (in Chinese)

[37] JGJ 94–2008. Technical specification for building pile foundation. Beijing: China Building Industry Press; 2008 (in Chinese)

[38] GB 50007–2011. Code for design of building foundations. Beijing: China Building Industry Press; 2011 (in Chinese)

Chapter 4

Structural Health Monitoring of Existing Reinforced Cement Concrete Buildings and Bridge Using Nondestructive Evaluation with Repair Methodology

Aman Kumar, Jasvir Singh Rattan, Nishant Raj Kapoor, Ajay Kumar and Rahul Kumar

Abstract

Sustainable development means the utilization of resources at a rate less than the rate at which they are renewing. In India infrastructure industry is growing rapidly due to globalization and raising awareness. In the present study, challenges faced by countries like India are to sustain the existing expectations with limited resources available. Reinforced Concrete (RC) structure may suffer several types of defects that may jeopardize their service life. This chapter deals with condition assessment and repair of RCC (G+3) building situated at Northern part of the country. There are various techniques available for repair and rehabilitation of reinforced concrete structures. From a maintenance point of view, it is essential to take up the strength assessment of an existing structure. So, to find out the reason behind the deterioration of the concrete structures some of the NDT and partially destructive technique are used. The NDT tests conducted during this study are Rebound Hammer, Ultra sonic Pulse Velocity, Concrete resistivity Meter, Ferro-scanning and Carbonation, etc. This chapter helps to explains, how identified the different parameters of distress building like strength, density, level of corrosion and amount of reinforcement. On basis of these results, apply a repair methodology to revert back the strength parameters of the buildings.

Keywords: visual inspection, nondestructive testing, repair, rehabilitation, bridges

1. Introduction

The major requirements in structures are to resist the deterioration due to aging. The repairing of the structures is very expensive [1]. In order to prevent the structure from deterioration, it is important to do a routine assessment of structures without damaging the internal part of the structure. The most common sort properties of concrete are its compressive strength, homogeneity of concrete along with corrosion probability [2]. This mentoring is only possible with nondestructive methods. The nondestructive test (NDT) methods are very useful in carrying out the in-situ condition of reinforced concrete (RCC) structures along with parameters causing deterioration of RCC structures [3]. In the present study, a case is carried out to access the levels of deterioration in an RCC structure located in northern India. Monitoring was carried out through detailed visual inspection along with NDT techniques to assess the deterioration for further planning of repair and rehabilitation. Many researchers have used nondestructive methods such as Rebound hammer and ultrasonic pulse velocity (UPV) to monitor the extent of deterioration in RCC structures [4]. The present pH value of concrete is determined with the help of phenolphthalein indicator and corrosion analyzed with corrosion analyzer [5]. There is a lot of demand for repair of damaged buildings and rehabilitation of existing concrete structures. The common structural defects are cracks, spalling, corrosion, leakage, chloride and sulfate attack, carbonation, etc. [1]. If such defects are not solved at their initial stages, it will lead to serious damages to the structures [6].

NDT techniques such as Rebound Hammer, UPV, Carbonation Test, Rebar Locator Test, and Impact Echo Test from a practical standpoint with an experienced Structural Engineer along with some partial-destructive testing methods are sufficient to access the quality and strength of in-situ concrete [3]. The NDT methods indirectly estimate the quality and strength of RCC structures, and the estimated results can be compared with destructive test results. The author suggested performing more than one test for better and accurate results [7]. The maintenance and utilization of new techniques/materials for repair/restoration of the buildings/structures are needed, for long-term sustainable development, especially, in developing country such as India. The author has discussed the various causes of deterioration and the methods for the repair, rehabilitation, and retrofitting [8]. The author also explained various materials and techniques for the repair, rehabilitation, and retrofitting and also the methodologies for the same [9]. The existing RCC buildings in Gujarat are located on the boundary of Gujarat–Maharashtra such as Nasik, Dhule, and Nandoorbar. In this study, the author has taken the case of a health building in the heart of the Nasik city. Based on the physical and experimental investigations, it was concluded that the structure either should be demolished or at least should be rehabilitated/retrofitted with appropriate method to enhance the service life of the building [10]. The RCC columns were strengthened by jacketing technique as this technique was more feasible and easy to execute at the site [11, 12]. All the columns on both the floors are now properly strengthened by jacketing, the concrete from the faulty slabs were completely removed, and the corroded reinforcement was changed with new reinforcement bars as per the design. And the slabs were recasted with M25 grade of ready-mix concrete [13]. In addition, selected tests and feasible techniques as per the latest advances in the industry to be used for health assessment, retrofitting, and rehabilitation are presented in depth with example and calculations [14, 15]. The objectives of the study are to suggest a model for a systematic approach for the repair of RCC structures. From the literature review, it is clear that RCC building undergoes deterioration and subsequent failure due to many reasons. It is, therefore, proposed to undertake a study on RCC framed buildings so as to suggest a proper methodology [16]. The building under consideration has been assessed for the causes of deterioration and for the repair, rehabilitation/strengthening work based on an investigation in order to suggest a suitable model for the rehabilitation of RCC buildings [17]. The outline of this chapter includes determination of building parameters such as strength, covercrete, density, level of corrosion, amount of reinforcement, pH value of concrete, and its carbonation depth.

The repair methodology for different structural members, deterioration types of concrete bridges, testing methods are needed.

3D printing concrete and digital fabricated techniques are on the high peak. The structural modeling of the 3D printing concrete is based on number of time-dependent mechanical properties, which include Young's modulus and compressive strength. These properties should be checked using destructive test methods is a time-consuming process. To avoid that, UPV waves were used to identified the density, strength, and Young's modulus of the concrete. Some of the tests in this chapter might be useful for this type of practices in the future [18].

RCC building located at the northern part of the India was selected to carry out the investigation. The building was around 32–35 years old. Detailed visual inspection was carried out to scrutinize the type, extent, and source for damage. An investigation was carried out to check the concrete quality, corrosion in reinforcing bars, carbonation of concrete, and ingress of salts in concrete. The concrete quality was found out by using UPV and hammer rebound method. The detailed investigation plan is shown in **Figure 1**.

2. Visual investigation

a. From the investigation, it was observed that the exterior columns of the building were having vertical cracks and spalling at some of the locations.

b. Dampness and efflorescence on walls were also observed, especially at the lower level of the building.

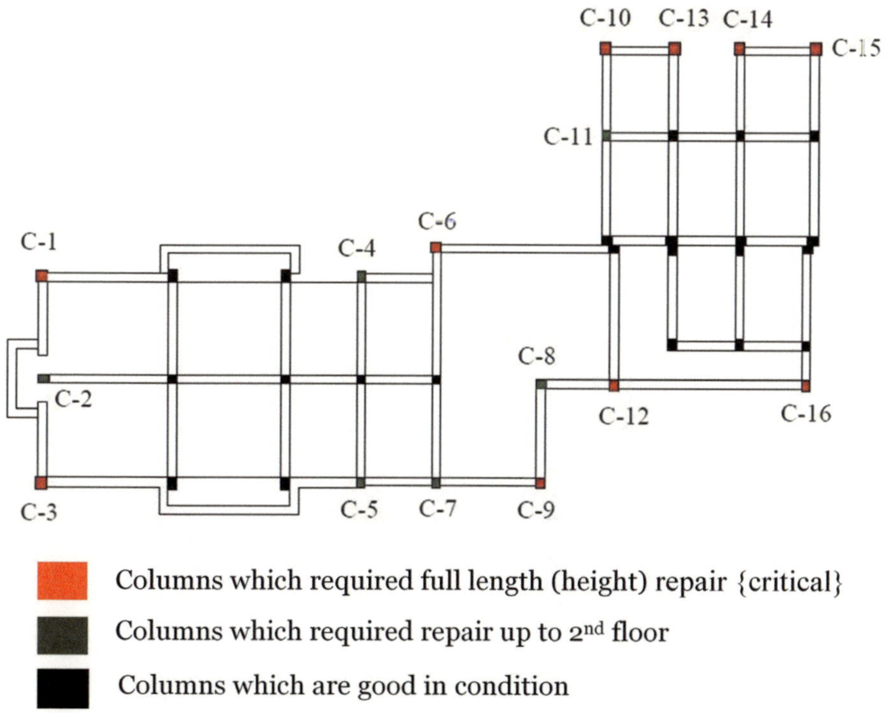

Figure 1.
Plan of the investigated building (not to scale).

c. Some minor cracks along the openings of the windows and doors were present.

d. Damaged staircase of servant quarter entrance from the backyard of the building was present.

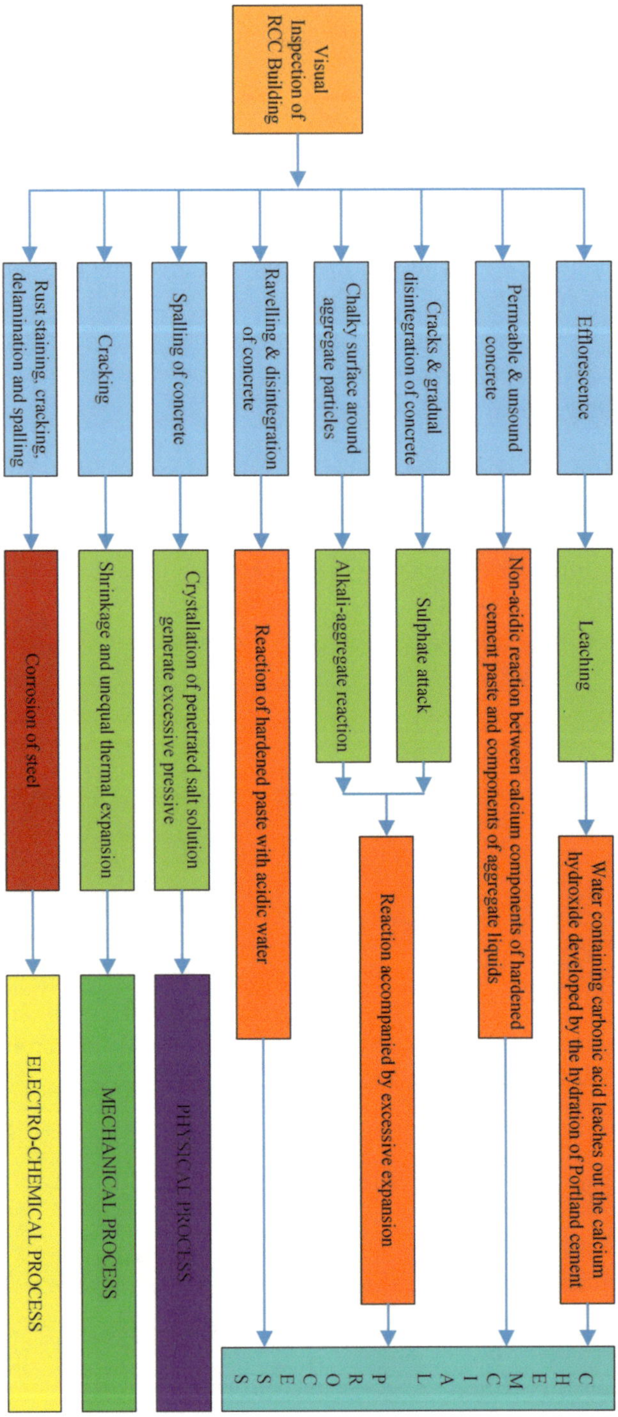

Figure 2.
Flow chart of visual inspection.

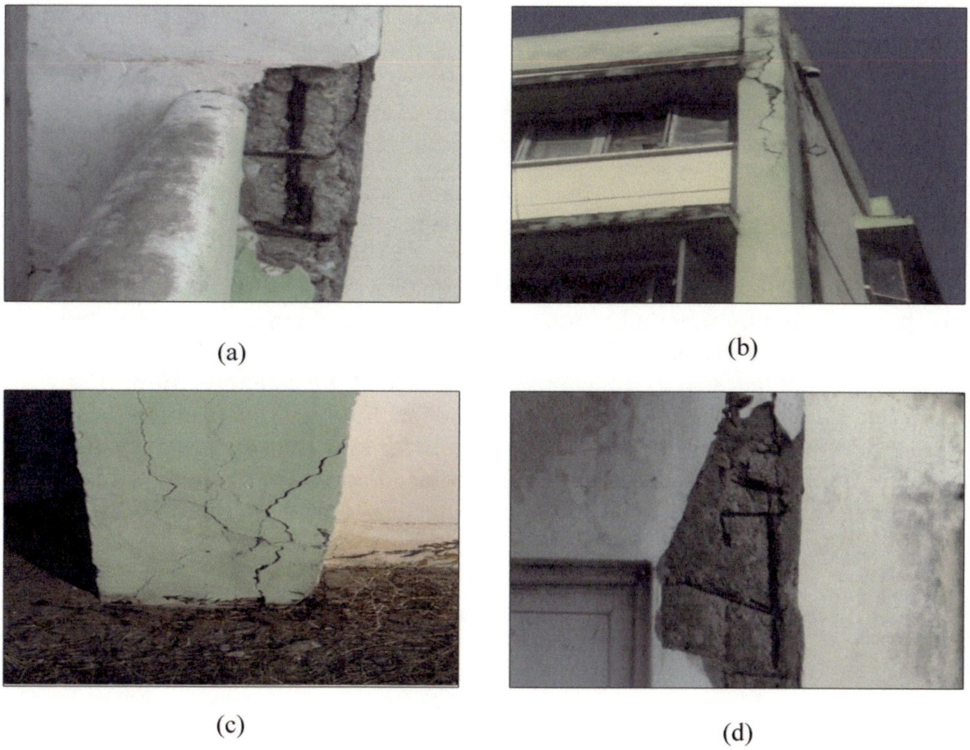

Figure 3.
Damage detection (a) spalling of concrete, (b) vertical crakes on column surface, (c) cracks in the bottom of column, (d) corrosion in steel.

e. Spalling of concrete and exposed corroded reinforcement in exterior columns of both the buildings.

f. Ground floor and the first floor of the building were attacked by termite.

g. Backside staircase along with the servant quarter of the building was completely damaged.

h. Brick tiles on the terrace of the building were damaged due to ingress of moisture through them.

i. Due to growth of plants on the terrace of the building and the rainwater pipes, there were damages.

The visual inspection checks found the presence of efflorescence, permeable, and unsound concrete, cracks, and gradual disintegration of concrete, spalling of concrete and cracking, etc. The flow chart of the visual inspection is shown in **Figure 2** [19–25]. Different types of distresses such as cracks, spalling, and corrosion in the building are shown in **Figure 3**.

3. Nondestructive evaluation

A number of nondestructive and partially destructive techniques for the assessment of the concrete structure are available to predict the cause of deterioration of the existing structure. These NDT techniques can be broadly classified into five groups such as strength tests, durability tests, performance tests, integrity tests, and chemical tests [2]. With the help of these tests, we can find out in-situ strength/quality of the concrete to precisely identify the damage and causes of the deterioration of the structure. Some of the most commonly used NDTs are discussed [21].

3.1 Rebound hammer (Schmidt hammer test)

This is the fastest method to evaluate the quality of concrete based on hardness, which is indicated by rebound number. If the strength of concrete is high, the rebound number is also high. The principle of this test is that when the plunger of the rebound hammer is pressed against the surface of the concrete, the spring controlled the mass rebounds and the extent of such rebound depends upon the surface hardness of the concrete. The surface hardness of the concrete gives the rebound number, which is further related to the compressive strength of the concrete. In latest rebound hammer, there is no need for angle correction. The average

Instrument	Average rebound number	Quality of concrete
Schmidt Hammer N- type	>40	Very good hard layer
	30–40	Good layer
	20–30	Fair
	<20	Poor concrete
	0	Delaminated

Table 1.
Rebound number with respect to quality of concrete.

Figure 4.
Rebound hammer.

value of rebound hammer for a different quality of concrete as per Indian Standard IS: 13311 Part-21992 is given in **Table 1** (**Figure 4**) [26].

3.2 Ultrasonic pulse velocity (UPV)

UPV Test method is generally used for determination of uniformity of concrete, to find crack depth, honeycombing, and to check the condition assessment of deterioration of concrete. The principle of this test is based on the propagation of electroacoustic pulse through the concrete pathway and then calculating the transit time taken, for a known distance. UPV mainly depends on the elastic modulus of the concrete. The general guidelines for quality of concrete as per Indian Standard IS: 13311 Part-11992 are given in **Table 2** (**Figure 5**) [27].

3.3 Concrete carbonation

Concrete has micropores, and these pores are filled with liquid, having pH values up to 12.5. Thus, the concrete is alkaline in nature. Carbonation of the concrete is the reaction of $Ca(OH)_2$ with the atmospheric CO_2 and its conversion into $CaCO_3$. This reaction decreases the pH value of the pore water up to 8.5. As time passes, the outer zone of concrete is affected first, and carbonation proceeds deeper into the mass as CO_2 diffuses inward from the surface. If the carbonation depth reached the depth of steel in concrete, then the steel is prone to corrosion damage. By carbonation test, we can measure the carbonation depth of the concrete. In order to determine the path of the carbonation, drilling a hole is done in stages and the phenolphthalein solution spread over it after every stage. As soon as the color of the concrete becomes pink, we stop the drilling process and the depth of the hole is measured (**Figure 6**) [28].

Sr. No.	Velocity (km/sec)	Concrete quality
1	>4.5	Excellent
2	4.0–4.5	Very good
3	3.5–4.0	Good to very good, slightly porosity may exist
4	3.0–3.5	Satisfactory but loss of integrity is suspected
5	<3.0	Poor and loss of integrity exists

Table 2.
General guidelines for concrete quality based on UPV.

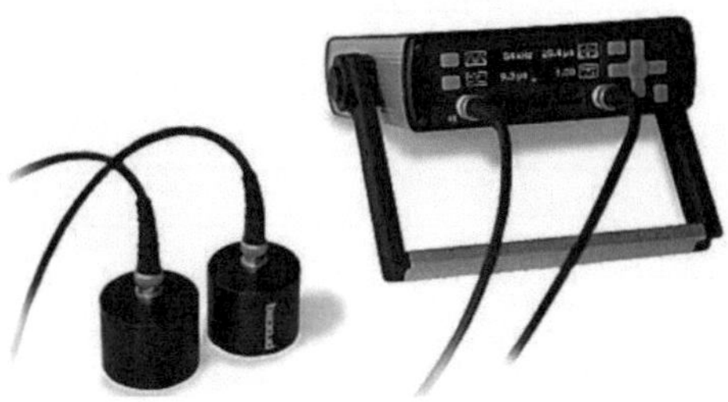

Figure 5.
UPV.

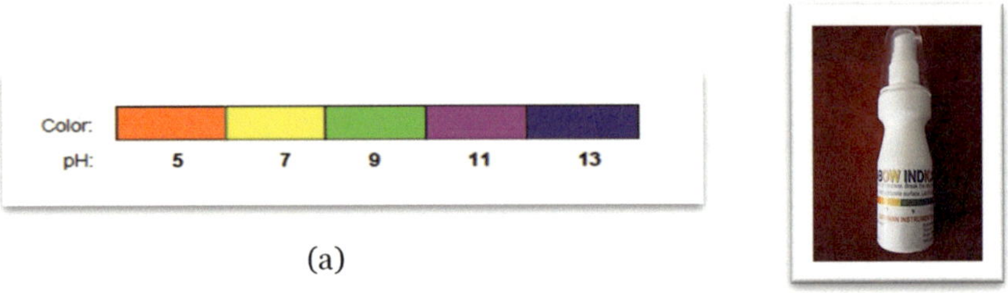

(a)

(b)

Figure 6.
Concrete carbonation: (a) pH scale for concrete carbonation. (b) Rainbow indicator.

3.4 Reinforcement scanning test

Ferro scanner is a device used to locate reinforcing bars and to estimate the diameter and depth of cover. This device is based on interactions between the bars and low-frequency electromagnetic fields. The ferro scanner works on the principle of electromagnetic induction, in which alternating magnetic field induces an electrical potential in an electrical circle intersected by the field. The test for reinforcement scanning is done with the help of HILTI PS 200 Ferro scan [29], a portable system for detecting rebar in concrete structures. The tools help in obtaining a real image of the reinforcement and evaluate the reinforcement mesh. HILTI PS 200 Ferro scan records the depth and positions of rebars over long stretches and obtains average coverage and statistics of the rebar diameter. The tools consist of image scanner that records the data, then the recorded data is transferred to the monitor for further analysis at the site itself. The major analysis is done on the computer by the analysis software to produce reports of the recorded data, and later the data is further submitted to the structural consultant for preparation of structural drawing, thereby ensuring the stability of the structure. The major limitations of this test are that the interferences may occur in images due to scraps of reinforcement in concrete, tie wires near rebars cross, and aggregates with ferromagnetic properties (**Figure 7**).

3.5 Concrete resistivity test

Surface resistivity meter provides very useful information about the surface of the concrete. It does not only provide corrosion information but also provides the corelation between the resistivity and chloride diffusion rate [30].

The operating principle of the Wenner probe, the Resipod, is designed to measure the electrical resistivity of concrete or rock. A current is applied to two outer

Figure 7.
Ferro-scanner.

probes, and the potential difference is measured between the two inner probes. The current is carried by ions in the pore liquid. The resistivity depends upon the spacing of the probes. The resistivity is obtained by the formula given below:

$$\text{Resistivity } \rho = \frac{2\pi a V}{I} \ [\text{k}\Omega\text{cm}] \tag{1}$$

where, a = distance between the probes as shown in **Figure 8** [30], V = potential measured, I = applied current.

The estimation of the likelihood of corrosion is given in **Table 3**.

3.6 Sulfate and chloride ingress test

Quantity of chlorides and sulfate in the concrete is generally determined chemically. Sulfate and chloride contents of concrete samples are collected from various locations. The permissible limit of chloride contents by weight of cement is 0.4%, and 0.15% is enough for the onset of corrosion. Sulfate content is limited to 4.1% by weight of cement [31].

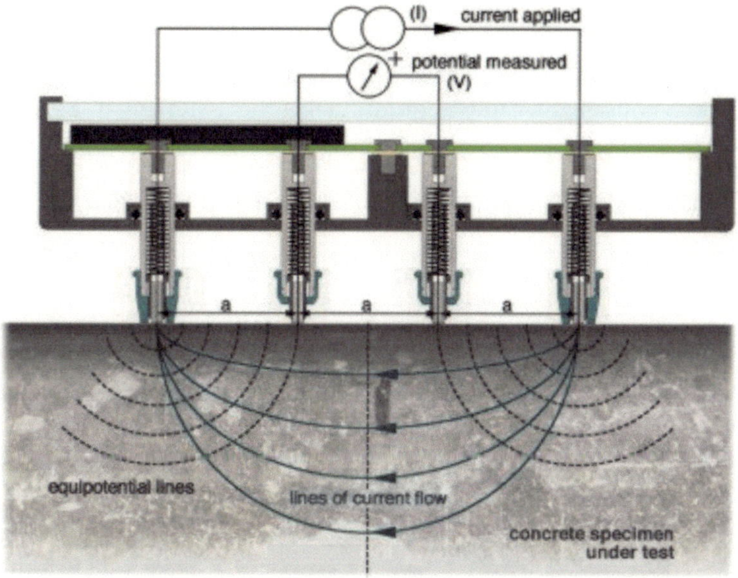

Figure 8.
Concrete resistivity meter.

Sr. No.	Resistively level (Kilo-ohm cm)	Possible corrosion rate
1	≥ 100	Negligible risk of corrosion
2	50–100	Low risk of corrosion
3	10–50	Moderate risk of corrosion
4	≤ 10	High risk of corrosion

Table 3.
The estimation of the likelihood of corrosion.

4. Test results of rebound hammer, UPV, extent of carbonation in concrete, level of corrosion, and chloride and sulfate content

The test results of rebound hammer with pulse velocity at different locations are tabulated in **Table 4**.

The test results of concrete resistivity to check the level of corrosion are given in **Table 5**.

The test results of chloride and sulfate content in concrete samples are tabulated in **Table 6**.

4.1 Summary of problems and defects

1. Cracks: Major cracks were observed at few locations in outer columns of the buildings. Minor cracks were observed near openings of windows and doors in most of the locations. A few cracks were also observed on the parapet of the terrace.

2. Rusting of bars: The corrosion was observed at few locations due to spalling of concrete or carbonation of concrete.

3. Spalling of concrete: Spalling of concrete is observed in outer columns of the buildings.

4. Dampness and efflorescence: Dampness and efflorescence were present in most of the houses especially at ground level, near sunken area, and near staircase areas.

5. Railing: The present railing is damaged and needs to be replaced.

Sr. No.	Id	RN	UPV	f_{ck}	f'_{ck}	G	pH
1	C-1	45.3	2010	24.5	22.05	25	6–7
2	C-2	43.4	1541	22	19.8	25	8–9
3	C-3	26.3	2073	10	9.00	25	6–7
4	C-4	44.7	1843	23.5	21.15	25	6–7
5	C-5	47	2226	26.5	23.85	25	6–7
6	C-6	41.3	1773	20	14.00	25	6–7
7	C-7	46.6	2833	26	18.20	25	6–7
8	C-8	28	2381	10.5	9.45	25	8–9
9	C-9	46.2	2030	25.5	17.85	25	6–7
10	C-10	44.8	2551	24	21.60	25	6–7
11	C-11	40.2	2212	19	17.10	25	6–7
12	C-12	30.9	1895	12	10.80	25	6–7
13	C-13	31.7	2368	14	12.60	25	6–7
14	C-14	36.2	2128	15.5	13.95	25	6–7
15	C-15	30.8	1531	12	10.80	25	6–7
16	C-16	32	2579	13	11.70	25	6–7

Table 4.
UPV, RH, compressive strength, and carbonation test result before repairs.

Sr. No.	Specimen Id	Concrete resistivity values (kΩcm)	Level of corrosion
1	C-1	42,44,38.2,34,36	38.84
2	C-2	18.1,16.4,15.7,15.9,13	15.82
3	C-3	23,27,20.6,20.4	22.75
4	C-4	42,37,36.2,33.1	37.07
5	C-5	42,44,38.2,34,36	38.84
6	C-6	21,22.4,16.8,17.8	19.5
7	C-7	23,27,20.6,20.4	22.75
8	C-8	22,23.4,25.6,28.5	24.88
9	C-9	18.1,16.4,15.7,15.9,13	15.82
10	C-10	34,42,44.3,41	40.32
11	C-11	18,18.5,22.3,22	20.2
12	C-12	21,22,21.8,23	21.95
13	C-13	24,26,20.4,20.6	22.75
14	C-14	45,43.2,42,32.5	40.67
15	C-15	11.4,16.2,14.3,13.9	13.95
16	C-16	28,36,31.8,34.2	32.5

Table 5.
Test results of concrete resistivity.

Sr. No.	Specimen Id	Chloride	Sulfate
		By weight of cement (%)	By weight of cement (%)
1	C-2	0.41	3.2
2	C-4	4.1	3.2
3	C-7	4.2	3.15
4	C-11	0.39	3.6
5	C-13	0.39	3.3
6	C-14	0.36	2.9

Table 6.
Sulfate and chloride content ingress.

4.2 Results and discussions

- From the UPV results, it was observed that all the structural members have very low UPV value. In total, 18.75% elements have the UPV value in the range of 2.5–3.0 Km/sec, 50% elements have the UPV value in the range of 2.0–2.5 Km/sec, and 31.25% members lie in the range of 1.5–2.0 Km/sec. All the UPV values lie below 3.0 (km/sec), which are doubtful. The UPV results cannot fulfill the codal requirements and are shown in **Figure 9**.

- The rebound hammer test results show that the compressive strength of concrete members with various range of percentages is 25%, 37.5%, 25%, and 25%. The compressive strength of these members lies in the range of 20–25 N/mm^2, 15–20 N/mm^2, 10–15 N/mm^2, and 5–10 N/mm^2 respectively and shown in **Figure 10**.

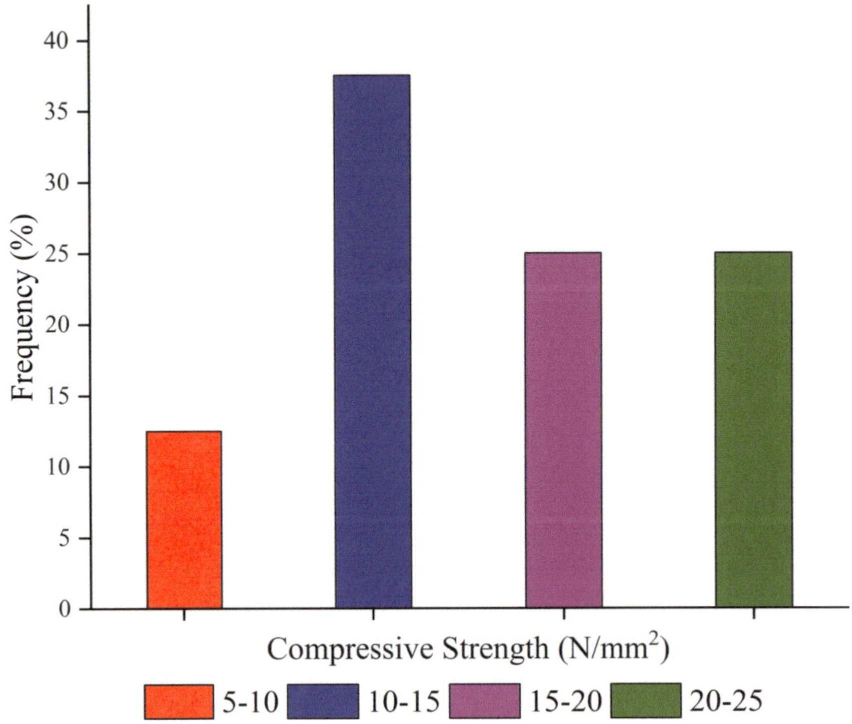

Figure 9.
Ultrasonic pulse velocity (km/sec).

- Adequate concrete cover over the reinforcement is one of the crucial parameters, as far as RCC structures are concerned, but the result of percentage distribution of concrete cover reveals that there is great variation in concrete cover. The specified cover as per drawing detailing is 40 mm. The percentage of less concrete cover is about 54.66% and shown in **Figure 11**. This can be one of the reasons for deterioration in concrete and subsequent corrosion in reinforcing bars.

- The pH value of external RCC members is low in the range of 6–7, and pH of internal members lies in the range of 8–9. From the carbonation depth data, it was observed that the depth of carbonation is more. In 45% of total members, the carbonation has reached up to the rebar level and is shown in **Figure 11**. This has been attributed to porous concrete and unprotected external surfaces from weathering actions.

- Concrete resistivity result showed that the members have a moderate risk of corrosion and is shown in **Figure 12**.

Observing the damaged condition of the outer columns, 10 columns marked in red color in the plan (**Figure 1**) require full height repair, and six columns marked in green color in the plan require repair/jacketing up to the second floor. Exposed concrete was found to be carbonated. The carbonated concrete should be provided with anti-carbonation coating if the spalling of cover concrete has not started. Due to the effect of corrosion, the spalling was observed in these columns, so, it is necessary to repair the structure so that it will be enabled to withstand against the combination of loads for which it is designed. The spalling concrete from columns should be repaired with micro-concrete. All the repair work should be carried out as

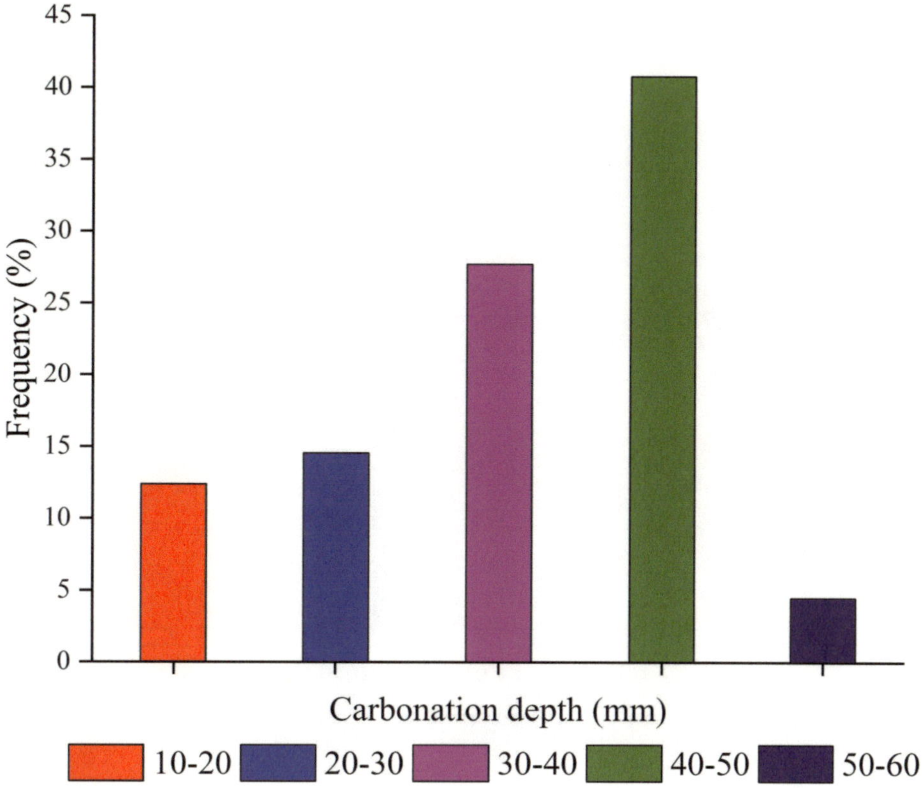

Figure 10.
Compressive strength of concrete members (N/mm²).

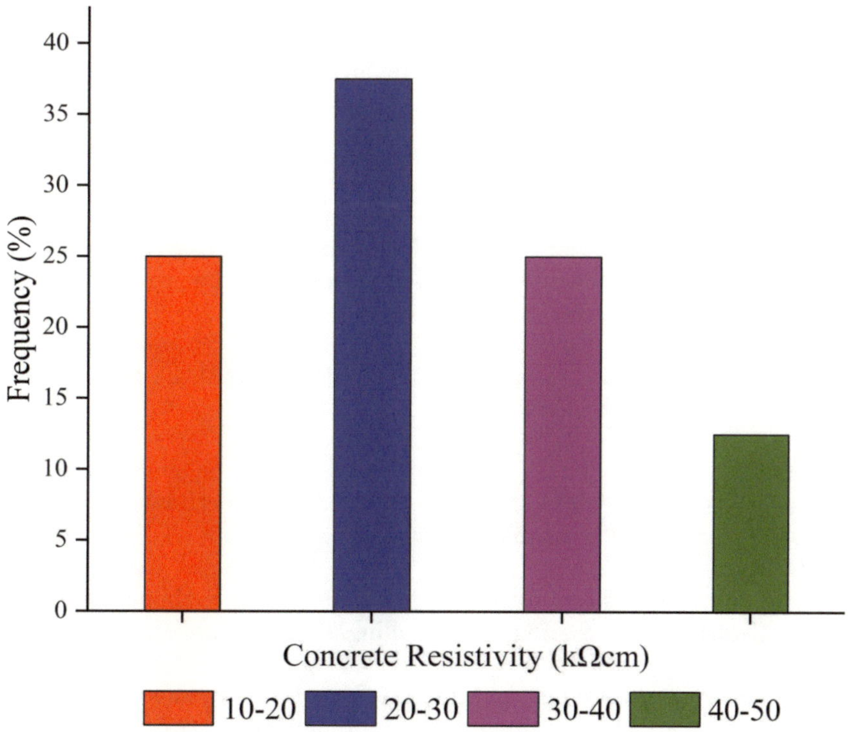

Figure 11.
Carbonation depth (mm).

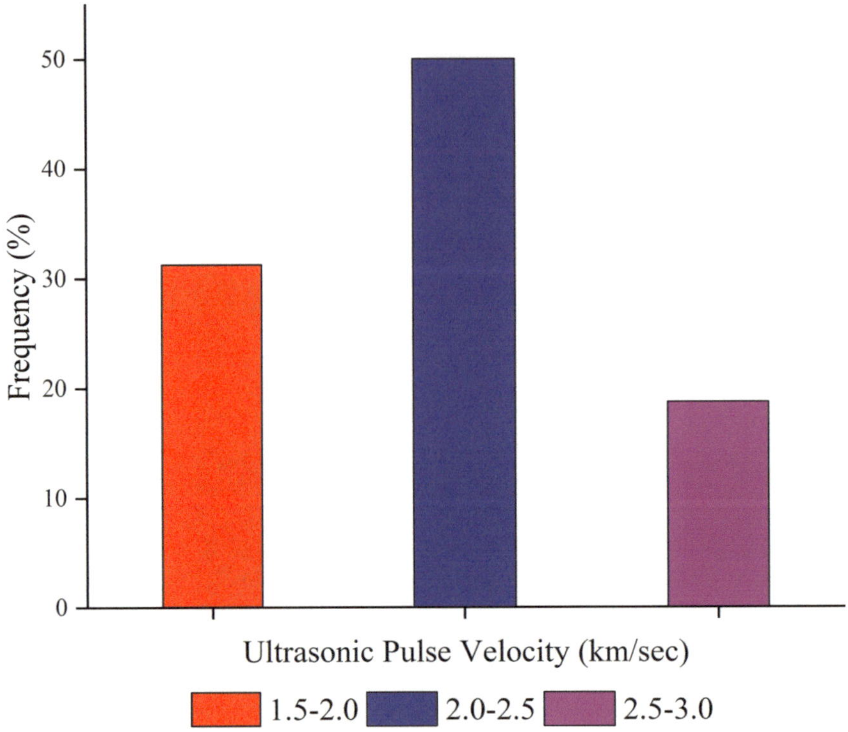

Figure 12.
Concrete resistivity (kΩcm).

per the methodology sequence provided above. The rebound hammer, UPV, and carbonation test results of concrete members before repairs are shown in **Table 4**. Concrete resistivity and chemical analysis results are in **Tables 5** and **6**, respectively.

4.3 Repair methodology

- The strengthening of outer columns of the building needs to be done with jacketing with micro-concrete, and reinforcement is to be provided where steel has rusted more than 20%.

- All the traps and manholes should be repaired to prevent the seepage into the foundations from such locations.

- Water tanks on the roof are causing dampness due to the overflow of water or due to leakage, all the tanks should be repaired, and overflow should be stopped by providing a suitable float valve.

- Exposed concrete was found to be carbonated and hence should be provided with anti-carbonation coating if the spalling of cover concrete has not started.

- If the spalling of cover concrete is taking place, the same should be repaired by treating the affected reinforcement and repairing the cover with micro-concrete.

The suggested model for carrying out repairs of structures and their strengthening for different types of problems is shown in **Figures 13–16**.

4.4 Test results of concrete columns after repair and strengthening

The NDT carried out after repair showed that the concrete strength was in the range of M25–M30 grade of concrete. The UPV also showed the quality of concrete improved from poor to good. **Figures 17** and **18** show that after repairs, the strength

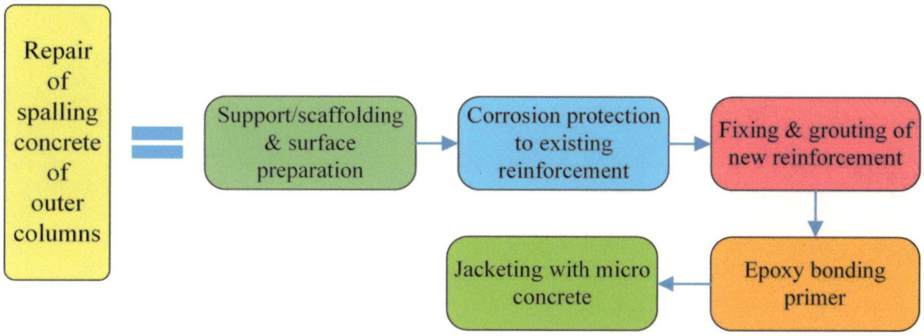

Figure 13.
Repair of spalling concrete of outer columns.

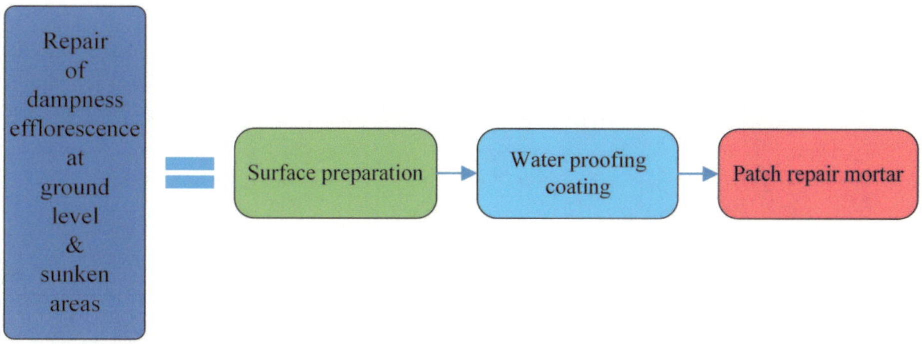

Figure 14.
Repair of dampness efflorescence at ground level and sunken areas.

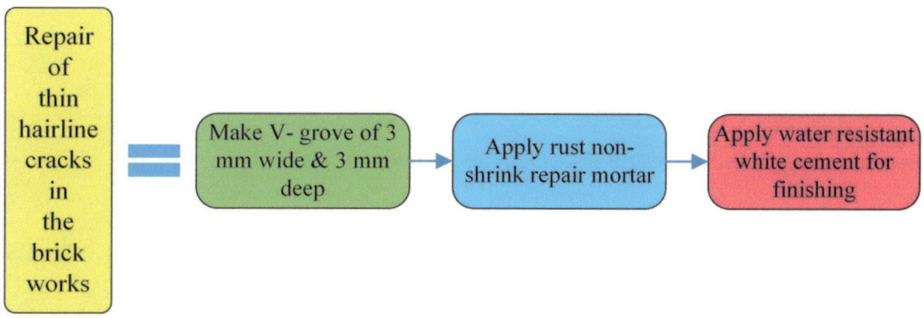

Figure 15.
Repair of thin hairline cracks in the brick works.

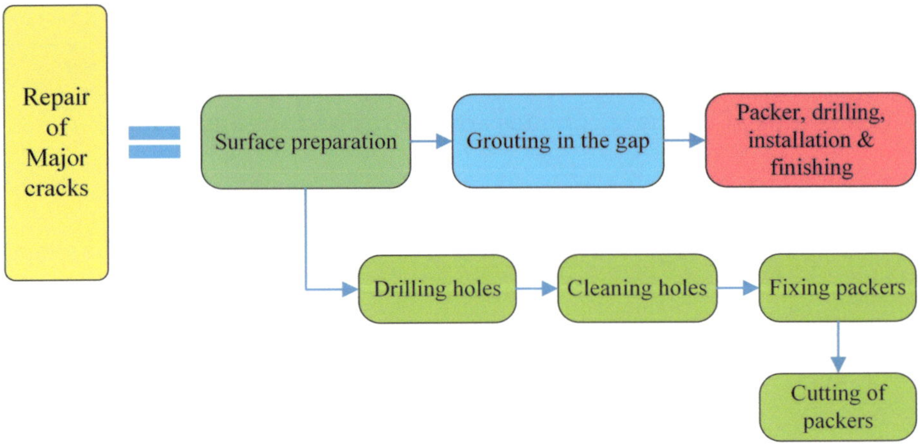

Figure 16.
Repair of major cracks.

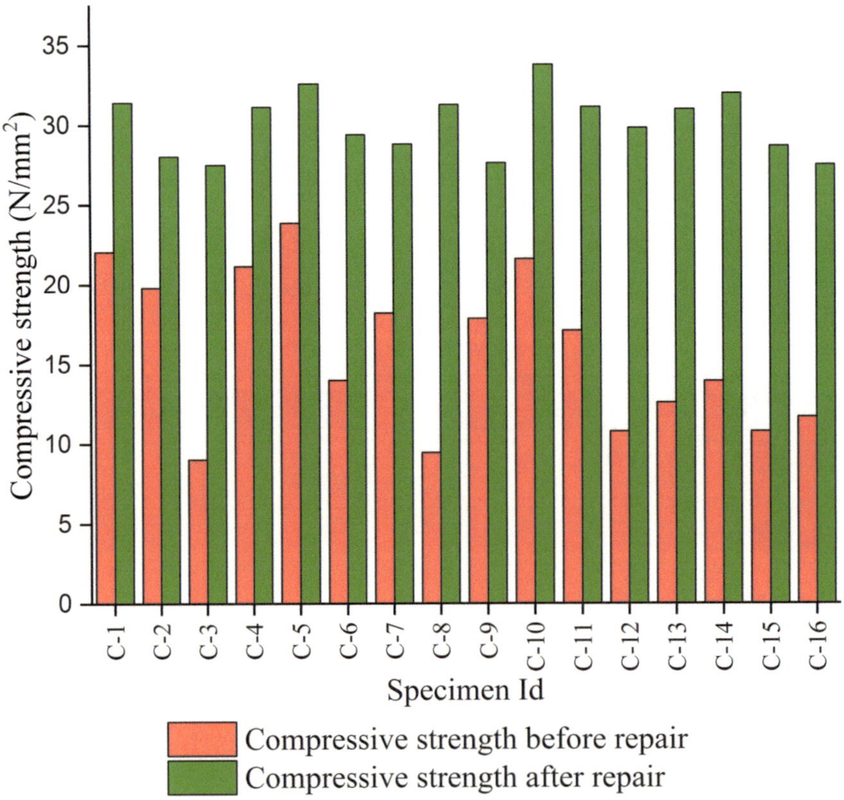

Figure 17.
Comparison of compressive strength before and after repair.

and quality were enhanced, which will also result in improvement of durability of structures in the long run.

Table 7 explains the test results of NDT data of repaired columns. All the enlisted columns are repaired with micro-concreting and in some locations, extra steel is also provided with anti-corrosion paint. This micro-concreting protects the

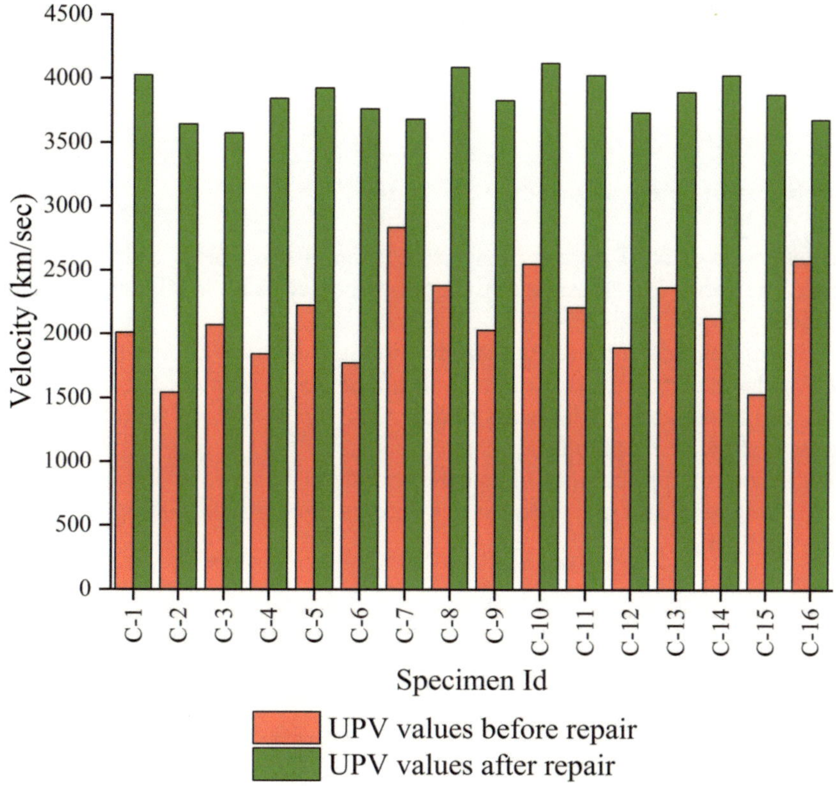

Figure 18.
Comparison of UPV values before and after repair.

Sr. No.	Id	RN	UPV	f_{ck}
1	C-1	50.6	4025	31.42
2	C-2	48.2	3643	28.00
3	C-3	47.8	3573	27.47
4	C-4	50.4	3844	31.13
5	C-5	51.4	3926	32.60
6	C-6	49.2	3763	29.40
7	C-7	48.8	3683	28.80
8	C-8	50.5	4088	31.27
9	C-9	47.9	3830	27.60
10	C-12	52.1	4121	33.78
11	C-11	50.45	4025	31.13
12	C-12	49.5	3735	29.81
13	C-13	50.3	3895	30.98
14	C-14	51	4025	32.00
15	C-15	48.7	3876	28.68
16	C-16	47.8	3680	27.47

Table 7.
Pulse velocity and compressive strength results after repair.

steel from corrosion because the density of this concrete is quite good. The anti-corrosion paint is helpful to protect the new attached steel as well as corroded steel from further corrosion.

5. Health monitoring of concrete bridges

The collection of NDT data in bridge structures is also similar as in the case of buildings. The condition assessment data of bridge deck slab is tabulated in **Table 8** and **Table 9**.

Activity of degradation mechanism dominantly depends on the type of material used for bridge constructions. Comparison of the importance of the basic chemical, physical, and biological mechanisms to deterioration of various materials of bridge structures is shown in **Table 8**.

Information on applicability of selected nondestructive load-independent techniques for detection and identification of defects during field testing of bridges is presented in **Table 9**.

Degradation mechanism		**Material of structures**					
		RC	**P_sC**	**St**	**M**	**T**	**S**
Physical	Accumulation of inorganic dirtiness	֍	֍	֍	֍	֍	֍
	Cyclic freeze–thaw action	֍	֍	֍	■	֍	
	Erosion	֍	֍	֍	■	֍	■
	Crystallization	֍	֍	֍		■	
	Extreme temperature/fire	■	■	■	֍	■	֍
	Creep	■	■	■		■	
	Relaxation	■	■	֍	■		
	Shrinkage	֍	֍	֍		■	
	Overloading	֍	֍	֍	֍	֍	
	Fatigue	■	■	■	֍	■	■
	Geotechnical condition changes	֍	֍	֍	֍	֍	
Chemical	Carbonation	֍	֍	֍		■	
	Corrosion	֍	֍	֍	֍		
	Aggressive compound action	֍	֍	֍	֍	֍	֍
	Chemical dissolving/leaching	֍	֍	■		֍	
	Reactions between material components	֍	֍	֍	֍	■	
Biological	Accumulation of organic dirtiness	֍	֍	֍	֍	֍	֍
	Activity of microbes	֍	֍	֍	֍	■	֍
	Activity of plants	■	■	■	■	֍	֍
	Activity of animals	■	■	■	֍	■	֍

Legend: ֍ - Basic Mechanism and ■ – Additional Mechanism.

Table 8.
Degradation mechanisms versus structural materials.

Technology	NDT technique	Class of defects					
		Df	DM	LM	Di	Co	Dt
Acoustic	Chain drag technique		■	■	■	■	
	Electromagnetic acoustic transducer		֍		■	■	
	Hammer sounding		■	■	֍		
	Impact eco		■	֍	֍		
	Impulse response		■	֍	֍		
	Parallel seismic				֍		
	Phased array ultrasonic		֍		֍		
	Time-of-flight diffraction		■	■	֍		
	Ultrasonic surface waves		֍	֍	֍		
	Ultrasonic tomography		■	֍	֍		
	Ultrasonic velocity		■	֍	֍		
Electrical and electrochemical	Electrical potential		■		֍		
	Electrical resistivity		■		֍		
	Microelectromechanical system	֍			֍		
Electromagnetic and magnetic	Alternating current field		֍		■		
	Eddy-current testing		■	■	֍		
	Electromagnetic conductivity		֍		֍		
	Magnetic flux leakage		■	֍	֍		
	Magnetic particle testing			֍	֍		
	Radar techniques	■	֍		֍		֍
Optic	Closed-circuit television	■				֍	■
	Geodesy/GPS surveying	֍	֍				֍
	Infrared thermograph testing		■	■	֍		
	Laser techniques	֍		■			֍
	Microscopy/endoscopy		֍	֍	֍	֍	
	Visual Inspection	֍	֍	֍	֍	֍	֍
Mechanical	Hardness testing		֍				
	Liquid penetrant				֍		
	Pressure techniques		֍				
	Sclerometric techniques		֍				
Radiological	Computer tomography	֍		֍	֍		
	Gamma or X-ray radiography		֍	֍	֍		
	X-ray fluorescence		֍			֍	
	Transmission radiometry		֍				

Legend: ֍ - Basic Technique and ■ – Additional Technique.

Table 9.
Defects detected by NDT techniques in filed testing of bridges.

Sr. No	Id	RN	UPV	f_{ck}	f'_{ck}	G	pH	R
1	S-1	46.00	3217	35.95	35.95	35	11–12	151.7
2	DS-2	46.17	3281	36.2	36.2	35	11–12	127.2
3	DS-3	45.75	3112	35.40	35.40	35	11–12	128.8
4	DS-4	46.25	3419	36.55	36.55	35	11–12	155.5
5	DS-5	46.08	3321	36.15	36.15	35	11–12	164.4
6	DS-6	46.08	3323	36.16	36.16	35	11–12	166.2
7	DS-7	46.91	2285	38.04	38.04	35	11–12	144.6
8	DS-8	46.75	3024	37.75	37.75	35	11–12	158.7
9	DS-9	46.00	2963	35.98	35.98	35	11–12	172.0
10	DS-10	46.75	3568	37.75	37.75	35	11–12	142.3
11	DS-11	45.83	3245	35.54	35.54	35	11–12	151.0
12	DS-12	46.25	3292	36.56	36.56	35	11–12	122.0
13	DS-13	46.33	3072	36.75	36.75	35	11–12	139.3
14	DS-14	46.25	3463	36.56	36.56	35	11–12	136.1
15	DS-15	46.17	3417	36.70	36.70	35	11–12	145.9
16	DS-16	45.83	3173	35.59	35.59	35	11–12	151.3
17	DS-17	45.66	3174	35.20	35.20	35	11–12	142.3
18	DS-18	45.92	3455	37.79	35.79	35	11–12	178.4
19	DS-19	47.5	3234	36.61	39.61	35	11–12	129.4
20	DS-20	46.167	3139	36.36	36.36	35	11–12	167.7
21	DS-21	46.67	3555	38.10	38.10	35	11–12	129.1
22	DS-22	45.58	3745	35.03	35.05	35	11–12	175.8
23	DS-23	46.08	3577	36.16	36.16	35	11–12	176.3
24	DS-24	45.92	3612	35.78	35.78	35	11–12	146.3
25	DS-25	45.6	3502	35.02	35.02	35	11–12	177.1
26	DS-26	46.1	3789	36.16	39.16	35	11–12	148.3

Table 10.
Test results of bridge deck slab.

The NDT parameters of bridge deck slab such as compressive strength, density, which is obtained from the UPV, pH of concrete, and level of corrosion are shown in **Table 10**.

There is no carbonation present in the concrete, so that the compressive strength and corrected compressive strength after carbonation are the same. The UPV values lie in the range of medium to good. Steel is free from the corrosion as the data given by concrete resistivity meter. The compressive strength of the deck slab varies in the range of 35–39 N/mm^2. The pH value of concrete is also in the normal range. **Figures 19–22** show the UPV values of deck slab (km/sec), compressive strength of deck slab (N/mm^2), cover depth (mm), and concrete resistivity (kΩcm), respectively.

6. Conclusion

In this chapter, the studied tests are rebound hammer, UPV, carbonation of concrete, concrete resistivity, covercrete, and ferrro scan. The repair methodologies

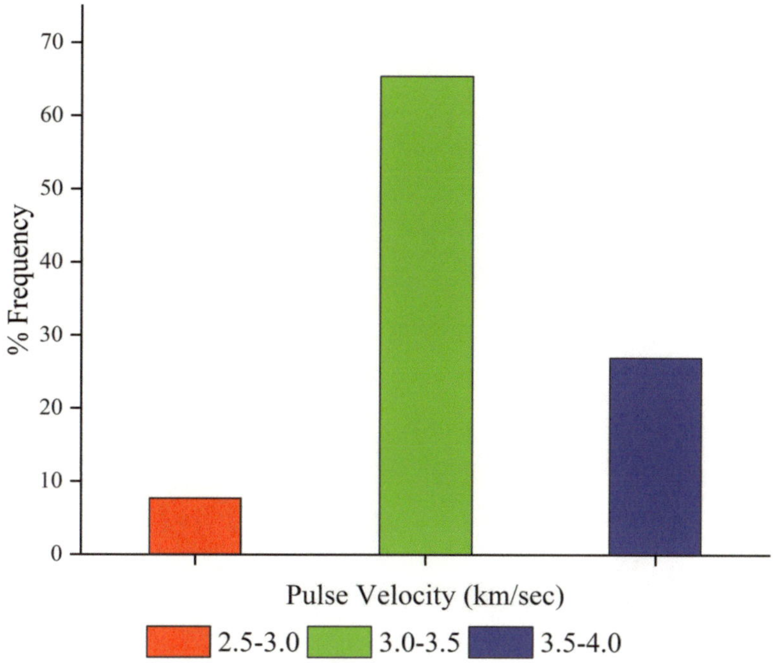

Figure 19.
Pulse velocity (km/sec).

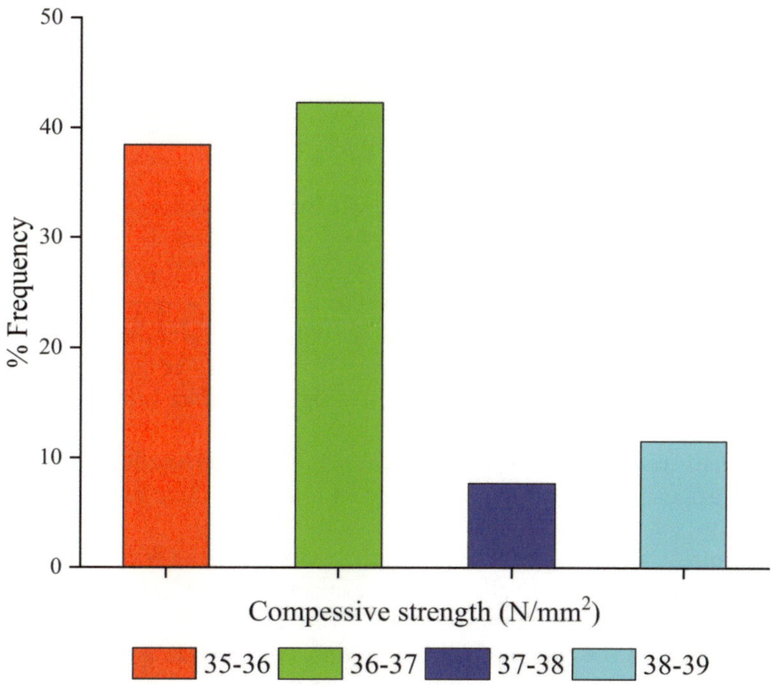

Figure 20.
Compressive strength (N/mm²).

are helpful to gain the mechanical properties such as strength and density of the concrete members more than original state of the structure. Based on the investigation, the following conclusions are drawn:

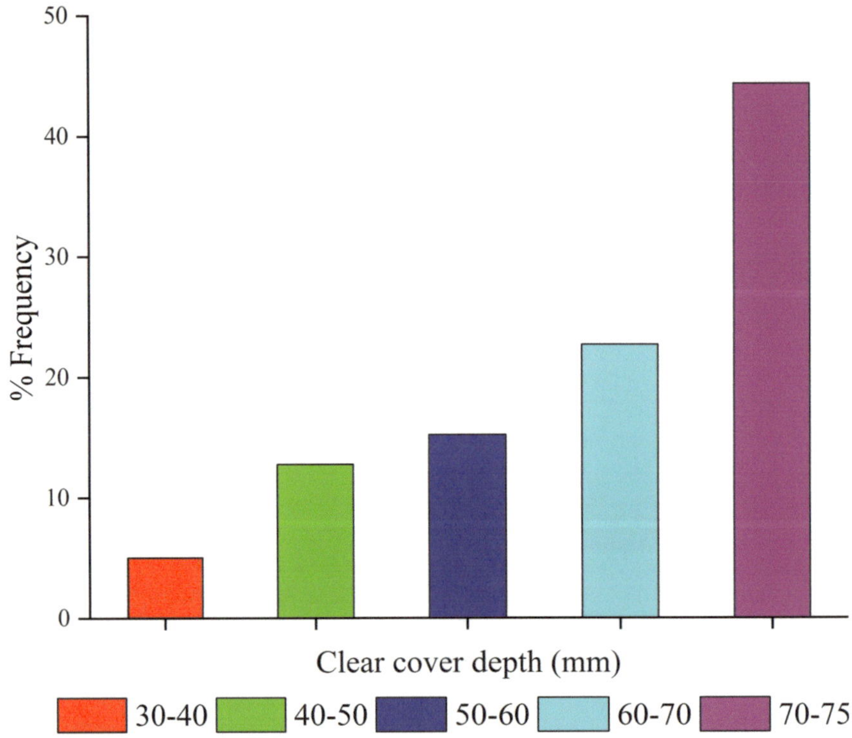

Figure 21.
Cover depth (mm).

- Rebound hammer and UPV test give physical properties of the in-situ RCC members.
- Deterioration of external exposed concrete members due to carbonation is high as compared with internal members.
- Adequate concrete cover to reinforcement is one of the important parameters for RCC structures. A major reason for spalling and corrosion in reinforcing bars is due to inadequate concrete cover. The porosity of concrete and unprotected external surfaces results in a high rate of carbonation.
- A systematic approach to repair and rehabilitation is to be adopted after carrying out the proper monitoring with the help of nondestructive testing techniques.
- The compressive strength of unstrengthened specimens with a frequency of testing specimens is as follows; 12.5% specimens are in the range of 5–10 N/mm^2, 37.5% specimens are in the range of 10–15 N/mm^2, 25% specimens are in the range of 15–20 N/mm^2; and only 25% of the specimens achieved designed grade of concrete, which is 25 N/mm^2. After strengthening all the specimens gained the strength more than 25 N/mm^2, data obtained from rebound hammer. The maximum obtained compressive strength was 33.78 N/mm^2 for specimen having Id C-12.
- The UPV of the test results shows poor quality of concrete, but after strengthening all the values lie in the very good to excellent zone.

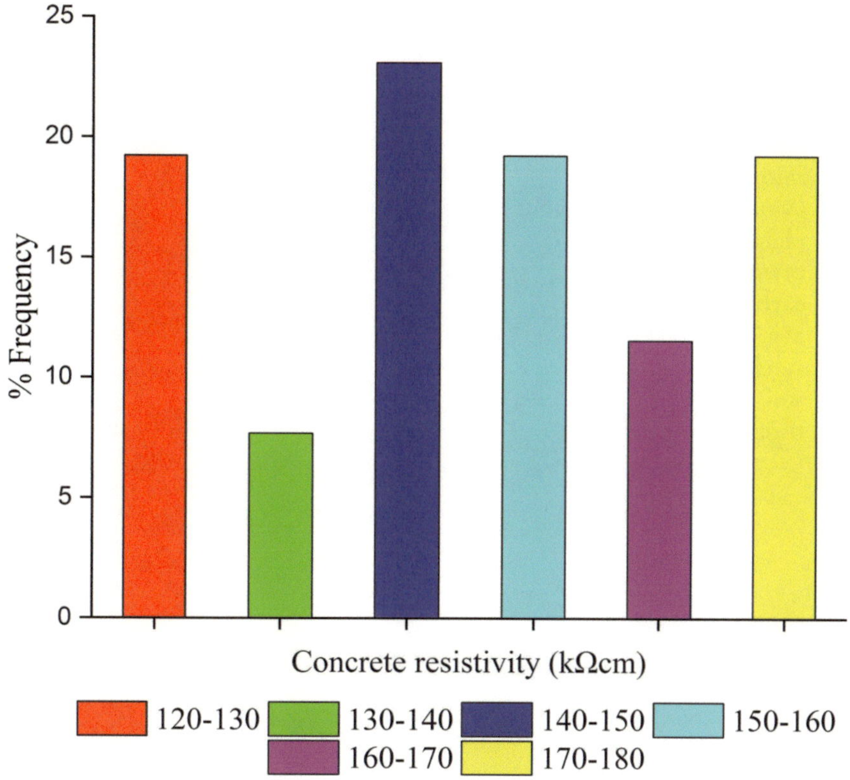

Figure 22.
Concrete resistivity (kΩcm).

- The monitoring to check the effectiveness of strengthening structures can be checked by NDT technique.
- The quality of concrete and steel in the brick deck slab is good and as per the designed detailing.
- There is further need to predict the nondestructive testing data using artificial intelligence.

Conflict of interest

The authors declare no conflict of interest.

Appendices and nomenclature

Co	contamination
Df	deformation
Di	discontinuity
DM	destruction of material
Dt	displacement
DS	deck Slab
f_{ck}	compressive strength (N/mm^2)

f'_{ck}	corrected compressive strength after carbonation (N/mm^2)
Id	specimen Id
LM	loss of material
M	masonry
G	grade of concrete
RC	reinforced concrete
RN	average rebound number
PC	plain concrete
P_sC	prestressed concrete
pH	carbonation range
UPV	avg. UPV (m/s)
St	steel
S	soil
T	timber

Author details

Aman Kumar[1,2*], Jasvir Singh Rattan[3], Nishant Raj Kapoor[1,4], Ajay Kumar[5] and Rahul Kumar[6*]

1 AcSIR - Academy of Scientific and Innovative Research, Ghaziabad, India

2 Department of Structural Engineering, CSIR-Central Building Research Institute, Roorkee, India

3 National Institute of Technical Teachers Training and Research, Chandigarh, India

4 Department of Architecture and Planning, CSIR-Central Building Research Institute, Roorkee, India

5 Gaur Associates, Haryana, India

6 KCC Institute of Technology and Management, Greater Noida, Uttar Pradesh, India

*Address all correspondence to: aman.civil16@outlook.com; rahulnitttr18@gmail.com

References

[1] Kovler K, Wang F, Muravin B. Testing of concrete by rebound method: Leeb versus Schmidt hammers. Materials and Structures. 2018;**51**:1-14. DOI: 10.1617/s11527-018-1265-1

[2] Brozovsky J. High-strength concrete —NDT with rebound hammer: Influence of aggregate on test results. Nondestructive Testing and Evaluation. 2014;**29**(3):255-268. DOI: 10.1080/10589759.2014.926897

[3] Jiang H, Han J, Li Y, Yilmaz E, Sun Q, Liu J. Relationship between ultrasonic pulse velocity and uniaxial compressive strength for cemented paste backfill with alkali-activated slag. Nondestructive Testing and Evaluation. 2020;**35**:359-377. DOI: 10.1080/10589759.2019.1679140

[4] Hobbs B. Ultrasonic Nde for assessing the quality of structural brickwork. Nondestructive Testing and Evaluation. 1995;**12**(1):75-85. DOI: 10.1080/10589759508952836

[5] Cawley P, Khalili P. The choice of ultrasonic inspection method for the detection of corrosion at inaccessible locations. NDT and E International. 2018;**99**:80-92. DOI: 10.1016/j.ndteint.2018.06.003

[6] Liu L, Miramini S, Hajimihammadi A. Characterising fundamental properties of foam concrete with a non-destructive technique. Nondestructive Testing and Evaluation. 2019;**34**(1):54-69. DOI: 10.1080/10589759.2018.1525378

[7] Kumar GB. Non-destructive testing in distress structure—An overview. Journal of Basic and Applied Engineering Research. 2014;**1**:114-119

[8] Puinotti R. Reinforced concrete structure: Non destructive in situ strength assessment of concrete. Construction and Building Materials. 2015;**75**:331-341. DOI: 10.1016/j.conbuildmat.2014.11.023

[9] Jagtap N, Mehetre P. Study on retrofitted R.C.C. building by different NDT methods. ISOR Journal of Mechanical and Civil Engineering. 2015;**12**(3):85-89. DOI: 10.9790/1684-12318589

[10] Scanlon A. Strength evaluation of an existing concrete bridge based on core and nondestructive test data. Canadian Journal of Civil Engineering. 2011;**14**:145-154. DOI: 10.1139/l87-026

[11] Verma S, Bhadauria S, Akhtar S. Monitoring corrosion of steel bars in reinforced concrete structures. The Scientific World Journal. 2014;**2014**:957904. DOI: 10.1155/2014/957904

[12] Subramanian N, Geetha K. Concrete cover for durable RC structures. The Indian Concrete Journal. 1997;**71**:197-201

[13] Hrushikesh S, Patil J, Dubey N. Damage detection and its repair and rehabilitation techniques by using NDT on RCC structures for corrosion mapping. International Journal of Advance Research in Science and Engineering. 2018;**7**(4):276-289

[14] Denys B, Balayssac J, Biondi S, Corbett D, Goncalves A, Grantham M, et al. Recommendation of RILEM TC249-ISC on non destructive in situ strength assessment of concrete. Materials and Structures. 2019;**52**:1-21. DOI: 10.1617/s11527-019-1369-2

[15] Kaur P, Singh J. Repair and rehabilitation of RCC structures. International Journal of Innovations in Engineering and Technology. 2020;**16**(4):14-20. DOI: 10.21172/ijiet.164.03

[16] Das S. Repair and rehabilitation of distressed structures including Indian case studies due to deterioration of

concrete and steel caused by environmental and other factors. International Journal of Engineering and Science & Technology.

[17] Lee S, Kalos N. Bridge inspection practices using non-destructive testing methods for concrete structure. Construction Research Congress. 2014; **2014**:1289-1298. DOI: 10.1061/9780784413517.132

[18] Wolfs R, Bos F, Salet T. Correlation between destructive compression tests and non-destructive ultrasonic measurements on early age 3D printed concrete. Construction and Building Materials. 2018;**181**:447-454. DOI: 10.1016/j.conbuildmat.2018.06.060

[19] Bhattacharjee J. Repair, rehabilitation & retrofitting of RCC for sustainable development with case studies. Civil Engineering and Urban Planning & An International Journal. 2016;**3**(2):33-47. DOI: 10.5121/civej.2016.3203

[20] Breccolotti M, Bonfigli M, Materazzi A. Influence of carbonation depth on concrete strength evaluation carried out using the SonReb method. NDT&E International. 2013;**59**:96-104. DOI: 10.1016/j.ndteint.2013.06.002

[21] Rakha T, Masri Y. A scoping review of non-destructive testing (NDT) techniques in building performance diagnostic inspections. Construction and Building Materials. 2020;**265**:1-12. DOI: 10.1016/j.conbuildmat.2020.120542

[22] Alcaino P, Maria H, Verdugo C, Lopez L. Experimental fast-assessment of post-fire residual strength of reinforced concrete frame building based on non-destructive tests. Construction and Building Materials. 2020;**234**:1-10. DOI: 10.1016/j.conbuildmat.2019.117371

[23] Hover K. Case studies of non-destructive test results and core strengths at age of 3-days. Construction and Building Materials. 2019;**227**:1-9. DOI: 10.1016/j.conbuildmat.2019.116672

[24] Huang Q, Gardoni P, Hurlebaus S. Adaptive reliability analysis of reinforced concrete bridges using nondestructive testing. In: First International Symposium on Uncertainty Modeling and Analysis and Management (ICVRAM 2011); and Fifth International Symposium on Uncertainty Modeling and Analysis (ISUMA). 2011. DOI: 10.1061/41170(400)112

[25] Cann D, Forde M. Review of NDT methods in the assessment of concrete and masonry structures. NDT&E International. 2001;**34**:71-84. DOI: 10.1016/S0963-8695(00)00032-3

[26] Indian Standard. IS 13311. Part-2. Non-Destructive Testing of Concrete-Rebound Hammer [Internet]. 1992. https://archive.org/details/gov.in.is.13311.2.1992/page/n5

[27] Indian Standard. IS 13311. Part-1. Non-Destructive Testing of Concrete-Ultrasonic Pulse Velocity [Internet]. 1992. Available from: https://archive.org/details/gov.in.is.13311.1.1992/page/n5

[28] ACI 562-13. Code Requirements for Evaluation, Repair and Rehabilitation of Concrete. American Concrete Institute. Available from: https://www.academia.edu/38287399/Repair_and_Rehabilitation_of_Concrete_Structures?auto=download

[29] PS 200 Ferroscan system. Hilti. Available from: https://www.hilti.in/c/CLS_MEA_TOOL_INSERT_7127/CLS_CONCRETE_SCANNERS_7127/r41791?itemCode=3623667

[30] Concrete Resistivity Meter. Available from: https://www.proceq.com/uploads/tx_proceqproductcms/import_data/files/Resipod_Sales%20Flyer_English_high.pdf

[31] Indian Standard. IS 456. Plain and Reinforced Concrete-Code of Practice [Internet]. 2000. Available from: https://archive.org/details/gov.in.is.456.2000/page/n3

Chapter 5

Behavior and Design of Transfer Slabs Subjected to Shear Wall Loads

Alonso Gómez-Bernal, Eduardo Arellano Méndez, Luis Ángel Quiroz-Guzmán, Hugón Juárez-García and Oscar González Cuevas

Abstract

This paper investigates the behavior of a transfer slab system used in medium rise building. For this purpose, two slab-wall full-scale specimens were designed, built, and tested to cyclic loads. The two slab-wall prototypes were exposed to three load stages: (a) vertical load, (b) horizontal load, and (c) vertical and horizontal combined load. The first specimen, SP1, includes a masonry wall situated on top of a squared two-way slab of 4.25 m by side, thickness of 12 cm, on four reinforced concrete girders, while the second specimen, SP2, consists of an identical slab but was constructed with a reinforced concrete wall. Some numerical finite element slab-wall models were built using linear and nonlinear models. The most important results presented herein are the change on lateral stiffness and resistance capacity of the load-bearing wall supported on a slab versus the wall supported on a fixed base and the effects that these walls cause on the slabs. During the experimental test process of horizontal loading, we detected that the stiffness of the two slab-wall systems decreased significantly compared to the one on the fixed base wall, a result supported by the numerical models. The models indicated suitable correlation and were used to conduct a detailed parametric study on various design configurations.

Keywords: transfer slabs, transfer floor system, discontinuity in buildings, shear walls on slabs, slab experimental behavior

1. Introduction

1.1 Buildings with transfer slab systems

In Mexican cities the construction of medium-rise buildings with a structural floor system called "transfer slabs" have been popularized in the last 15 years. A study [1] evaluated a set of buildings constructed in a sector of Mexico City, in this evaluation was detected a high percentage of buildings that were constructed with discontinuous walls, a structural configuration that induced irregularities in several buildings. These structures, projected with transfer floor systems represent a high risk, because it is known that buildings with discontinuity in elevation are vulnerable to seismic loads. This situation is critical when load-bearing walls, especially on the first floor, are not aligned with the vertical forces, therefore increasing seismic vulnerability.

The buildings designed with transfer slab (or transfer floor) have a floor system supported on one rigid level, which is used as a parking lot. On top of this transfer slab is constructed a shear masonry walls super-structure of more than four stories. A significant percentage of these walls are interrupted at transfer floor level and are discontinuous in the foundation. Nevertheless, a few walls in the boundaries of the structure are continuous until the base, but many walls in upper stories are not aligned with the framed axes of the bottom story. Numerical models have demonstrated that this structural configuration produces a significant increase in the shear stress in discontinuous walls, as is discussed in detail in Ref. [2]. This can be explained due to the excessive deflections that walls provoke to slab. Shear forces calculated are between two to three times larger than those that these walls would have if they were continuous along all their height. Besides, the transfer slab is exposed to additional high deformations and stresses.

During the September 19, 2017 earthquake (M = 7.1) an extensive damage was observed in Mexico City, some buildings structured with transfer slab system suffered important damage [3]. For example, **Figure 1a** show the total collapse of a building located in the Lake bed zone which had eight stories, the structure was located at the corner of a block, only the ground story consisted of reinforced concrete frames, while the upper stories were structured only with masonry walls conforming a structure with high discontinuity in elevation; this building collapsed due a high irregularity killing 5 residents. **Figure 1b** shows other two cases that correspond to two modern buildings structured with this transfer floor system, when these structures were reviewed, minor diagonal cracks were detected at the ends of the girders of the floor transfer (**Figure 1b**), in one of these buildings residents were evicted.

The structural system based on transfer floors requires a major research. In the literature exists a great number of experimental studies on the performance of load-bearing walls subjected to horizontal forces that has been characterized the seismic effects. Walls that have different parameters, in terms of materials, steel reinforcement, load conditions and other properties have been tested [4–7], and some of the most notable results were documented [8]. Nevertheless, in terms of performance of RC slabs, there is less information due to the limited number of experimental programs [9–11]. In these studies, the walls are fixed in their base and slabs were subjected to distributed or concentrated loads; however, there was no specific information about linear loads applied directly on the slabs, i.e. slabs supporting walls. The behavior between a building structured with transfer floors and one traditional

Figure 1.
(a) left: Collapse of a building with discontinuities in elevation; (b) right: Diagonal cracks in the girders of the transfer floor in two buildings supporting six and seven stories, respectively.

(without discontinuous walls) is different. For this reason, was implemented our experimental program.

1.2 Objective and scope

The objective of this study is to develop an experimental and a numerical program that considers the interaction between a wall and the slab on which is supported. In the experimental program, two full-scale slab-wall specimens were designed, assembled, and tested by cyclic loads in the Laboratory of Structures at UAM. The prototypes slab-wall were exposed to three load stages: vertical load, horizontal load, and combined vertical and horizontal load. The first specimen, SP1, include a masonry wall situated on top of a squared two-way slabs, while the second specimen, SP2, consists of an identical slab but was constructed with a reinforced concrete wall. To support the experimental results, and characterize the behavior of transfer floor structures, some numerical finite element slab-wall systems were analyzed using linear and nonlinear criteria.

2. Experimental program

2.1 Tested specimens, SP1 and SP2

Two slab-wall specimens were constructed in the experimental program. This program included three load stages: (a) vertical loads representing the vertical load component detected in buildings structured with floor system, (b) horizontal cyclic loads applied at low load levels in order to investigate the linear response, and (c) combined constant vertical and horizontal loading cycles.

Specimens SP1 and SP2 consist of four reinforced concrete perimetral girders with a solid two-way slab which together conform a monolithic structure. These girders have a transversal section of 25 × 70 cm, and the slab is square of 425 × 425 cm and 12 cm thick, the slab plane is located at 77 cm over the laboratory floor (see **Figure 2**). The concrete strength was f'c = 250 kg/cm^2. The two beams aligned transversals to the wall are haunched girders, while the two girder parallels to it have a rectangular section. The haunched girders were used to adjust the specimen to the conditions of the laboratory's reaction floor, and the position of the laboratory's horizontal load reaction system. The design and steel reinforcement in the slab were determined using typical procedures, such as the ones used in real buildings in Mexico. On the outer strips, the spacing of the bottom re-bars was set to 40 cm. On the central strip, the spacing was set to 20 cm in both directions. In the top of the slab, the reinforcement spacing was also set to 20 cm. **Figure 3** shows details of the reinforcement used in the slab specimens.

The first specimen, SP1, include a masonry wall placed on top of the slab. This wall is 2.50 m wide by 2.41 m high, was constructed on the slab's central strip. The wall is confined with two tie-columns, 14 × 18 cm, and with a tie-beam, 14 × 14 cm. The bottom tie-beam was integrated with the slab. The concrete resistance of the tie-columns was f'c = 150 kg/cm^2, with four re-bars in the corners. The re-bars are conformed with the minimum steel requirements of the Mexico Construction Code for masonry structures [12].

The second specimen, SP2 (**Figure 2**), consists of a slab with a Reinforced Concrete (RC) wall, reinforced with electrowelded wire mesh, besides at the ends were placed tie columns. The wall was placed on top of a squared two-way slab, with the same characteristics of SP1 specimen. The RC wall is 2.50 m wide by 2.40 m high, and with a thickness of 10 cm (top-right of **Figure 3**).

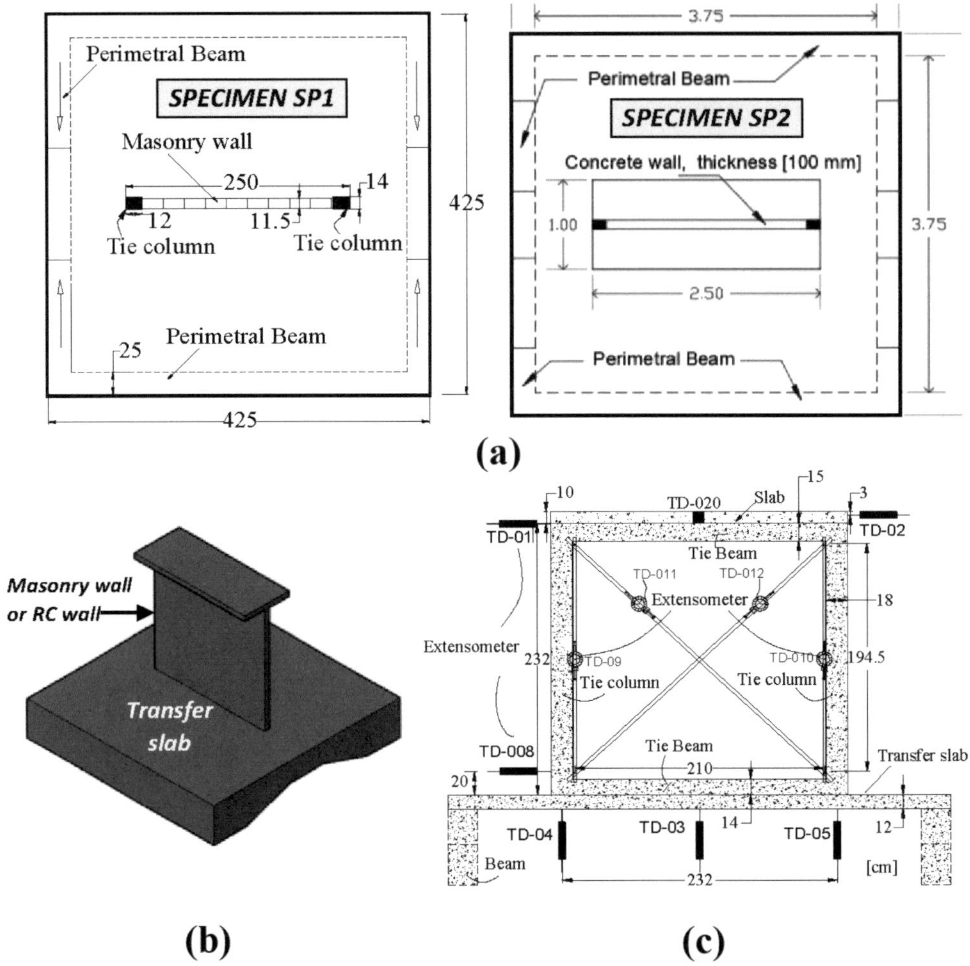

Figure 2.
Geometry and details of slab-wall specimens SP1 and SP2: (a) plan view of specimens: (b) transfer slab 3D model; (c) location of exterior instrumentation.

2.2 Setup, devices, instrumentation, and loading scheme

Three mechanical steel devices were designed and built to develop the experimental program: the first for attachment purposes, the second one for simulate the gravitational loading, and the third one for the lateral cyclic loading. The attachment device prevented the slab-wall specimen from moving in a vertical or lateral way during the test whenever the loads were applied to the wall. The vertical load device was fabricated by a system of steel girders, eight tensors and their anchorages (see **Figure 4**). At the top of the girders there are four vertical actuators (each with a 25 Ton capacity), supported on a base plate, which keeps them in an upright position and prevents them from slippage. The third device was designed for the incremental cyclic horizontal loading. We used a double-action hydraulic actuator with a 25 Ton push capacity, and a 13 Ton pull capacity. A more detailed description can be found in [13].

The interior instrumentation consists of a set of strain gages (SGs) that were installed in the reinforcement bars of the concrete elements. In specimen SP1, 16 SGs were fixed in the bars of the bottom of slab and 16 SGs in the top (**Figure 3**). The tie-columns of the masonry wall had 20 SGs in the longitudinal re-bars. Additionally, 10 SGs were embedded in the concrete slab and the tie-columns. In specimen SP2,

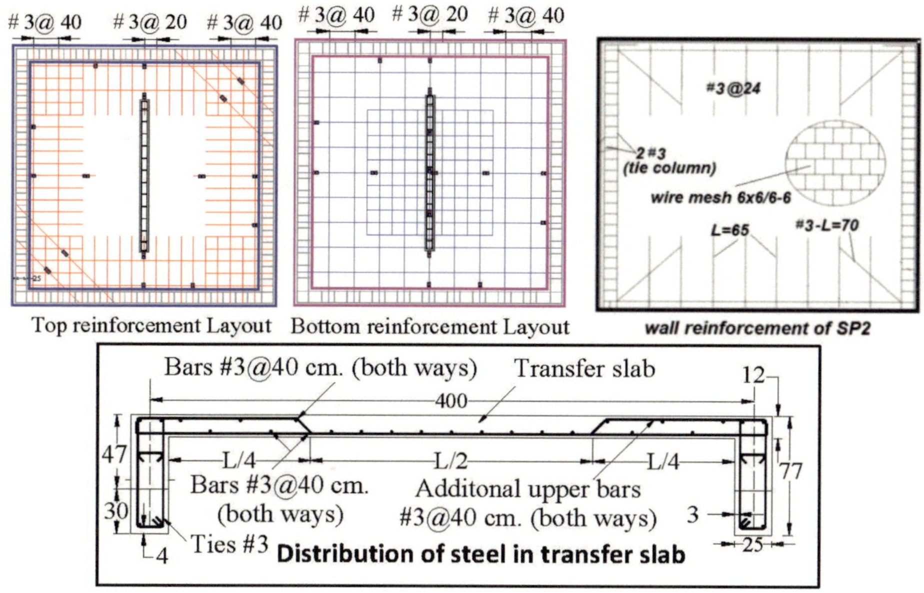

Figure 3.
(Top) Reinforcement details in transfer slab and reinforced concrete wall of SP2; (Bottom) assembly of strain-gages in bars is included.

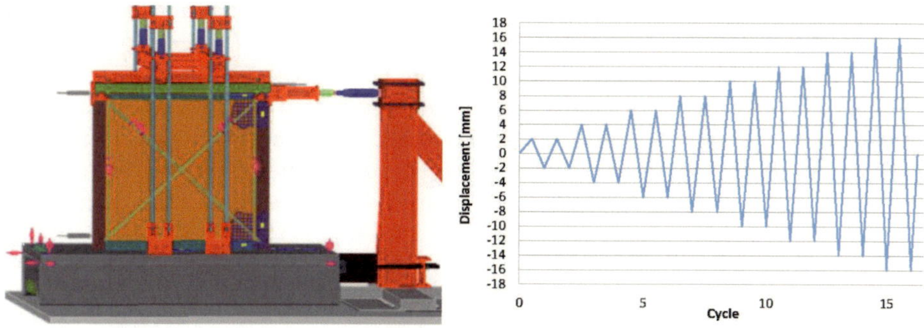

Figure 4.
Test set-up and typical displacement time history (horizontal loading).

a total of 48 SGs were fixed along the longitudinal and transversal directions of the reinforced bars.

The exterior instrumentation consisted of two parts, first, in order to measure the forces applied through the hydraulic cylinders, load cells (LVDT's) were attached (four verticals and one horizontal); and second, 20 transducers (TDs) were installed at strategic points with the purpose to measure the vertical and lateral deflections. **Figure 2c** shows the TDs on the lower face of the slab. The TD that provides the most relevant information is TD03 because it is in the center of the slab, exactly below the center of the wall (masonry in SP1, or concrete in SP2). The TD03 transducer is used as a control node to determine vertical displacements.

3. Numerical models

In a previous work [2], numerical inelastic models for the slab-wall system were studied using finite element (ANSYS [14]). Twelve detailed geometries were analyzed, six models were defined for wall masonry, and six systems for reinforced concrete wall. **Table 1** shows the geometric characteristics of six studied models, four different wall lengths and two slab thicknesses were used. The model was divided in two sections for reducing computational time in analysis. The support conditions of the slab are theoretically restricted along the perimeter. The models M10_M2.5_12V, M10_M2.5_12VL, M1_C2.5_12V and M9_C2.5_12VL are like the tested specimen.

All slab-wall numerical models were subjected to two load conditions, first to an incremental vertical load, and second to a combined load consisting of a simultaneous constant vertical component with incremental cyclic horizontal load. Capacity curves for the systems were generated with the obtained results. In addition, for comparison purposes, analysis of the vertical load condition included a slab model (without wall) subjected to a uniform load. For horizontal load analysis, a nonlinear model with wall on a rigid basis was used for comparison purposes (flexible base wall).

4. Definition of specimen response

In order to define force-deformation behavior and the different damage states of the two tested specimens, it is necessary to establish the deformation modes or configurations for each of the three load protocols studied (**Figure 5**). Therefore, the failure in the walls of the specimens, SP1 and SP2, will depend on the bending (ΔF) and shear (ΔC) deformations induced by horizontal loading. However, in this case, since each wall is supported by a slab, the lateral displacement due to the wall rotation (ΔL), induced by the slab deformation, must be considered. Total displacement, ΔT, at the top of the wall is then generated and presented in **Figure 5**.

Shear deformations of the walls can be obtained from recorded deformations of transducers installed in diagonal directions. Based on the strength of materials theory, the unitary angular deformation, γ is due to the shear stress acting on the wall element [4]. In general, this deformation is defined by the following expression:

$$\gamma = \frac{\delta_2 L_2 - \delta_1 L_1}{2 l_m h_m} \tag{1}$$

where γ = angular deformation of the wall, δ_1, δ_2 = shortening or elongation measured on the diagonals, L_1, L_2 = initial length of diagonals, l_m, h_m = are the width and height of the wall, respectively.

And the total distortion (R_{tot}) or effective distortion (R_{Eff}) is calculated as:

$$R = \Delta E / H \tag{2}$$

$$\Delta E = \Delta F + \Delta C - \Delta L \tag{3}$$

$$\Delta L = (\Delta I - \Delta D) \times H / L \tag{4}$$

where R = effective drift of the wall, ΔE = effective lateral displacement at top of the wall, ΔL, ΔF, and ΔC are defined in **Figure 5**, ΔI, ΔD are vertical displacements at ends in wall-slab joint, H = distance between TD-01 y TD-08 (**Figure 2**), L = distance between TD-04 y TD-05 (**Figure 2**).

Concrete wall model	Masonry wall model	Wall width (m)	Slab depth (cm)
M1_C2.5_12V	M2_M2.5_12V	2.50	12
M3_C3_12V	M4_M3_12V	3.00	12
M7_C3_13V	M7_M3_13V	3.00	13
M9_C2.5_12VL	M10_M2.5_12VL	2.50	12
M11_C3_12VL	M12_M3_12VL	3.00	12
M15_C3_13VL	M16_M3_13VL	3.00	13

Table 1.
Code for slab-wall numerical models [2].

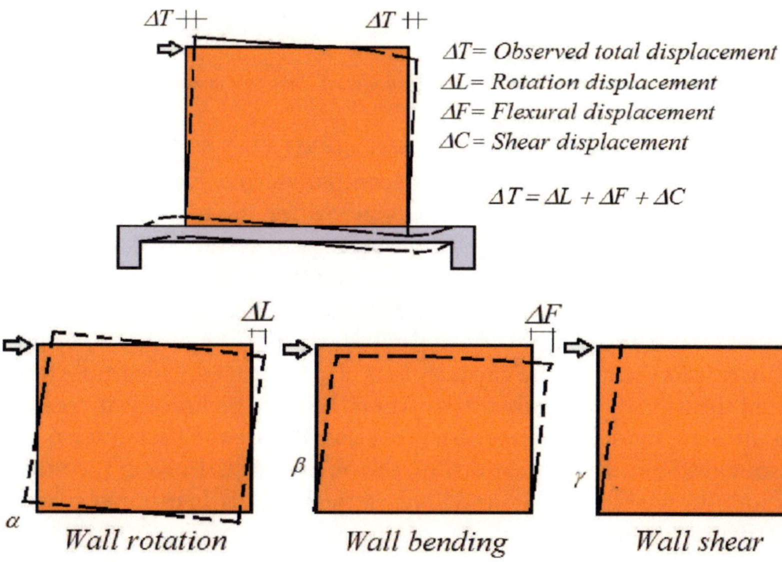

Figure 5.
Total deformation (above) and principal deformation modes.

5. Experimental and numerical results

5.1 Response to vertical loading, experimental vs. numerical results

In **Table 1** are defined the key to numerical models as: M#_Mat#_#V. The M# indicates the model number. The Mat# indicates the material and length of the wall, for example M3.0 indicates that the wall material is Masonry, and has a length of 3.0 m, while C indicates that the material wall is Reinforced Concrete. The #V indicates the slab thickness, for example 13 V indicates that the slab is 13 cm thick.

The experimental (Exp) capacity curve of SP1 is compared with the numerical model curve (M2_M2.5_12V). It can be observed that in the experimental case a minor stiffness value is obtained, as is observed in the curve. This situation can be explained due to the perfect fixed ends along the slab perimeter in the numerical model. But in both capacity curves, the first slab cracking occurred at the same time. This occurs when the load reached a value of 5.5 Ton (**Figure 6**). Numerical model, M2_M2.5_12V, was conducted until the failure condition, the first slab cracking take place at 5.5 Ton; the whole structure exhibited linear behavior until the ultimate load of 25 Ton. At this point an almost perfect elastoplastic "nonlinear" performance was detected (**Figure 6**).

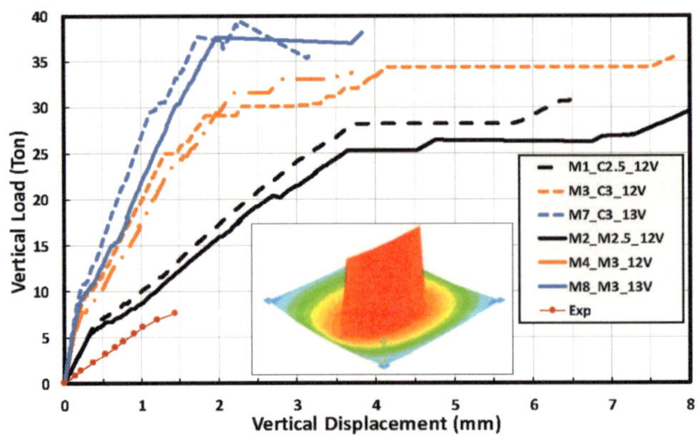

Figure 6.
Capacity curves obtained from numerical and experimental models, subjected to vertical loading.

When the capacity curves of numerical models M1_C2.5_V12 (reinforced concrete) and M2_M2.5_12V (masonry) are compared, they have a similar trend. This behavior indicates that under gravitational loading, no matter what material the wall is made of, the capacity of the whole structural system will almost remain the same. This result is corroborated with the other two pairs of curves shown in **Figure 6** (for example M3_C3_V12 and M4_M3_V12), which also described the same trend. When the wall length or slab thickness (M7_C2.5_V13 and M8_M2.5_V13), were increased the ultimate capacity was also increased. One important result obtained from the numerical study is that when the slab thickness is increased in models, the ultimate model capacity also increases. Before the first crack occurs, it can be assumed to be a service condition, the observed stiffness in the models during this condition were similar, and hence a thicker slab does not ensure a significant stiffness value.

5.2 Response to vertical and lateral loading

Specimens SP1 and SP2 were exposed to a constant vertical load of 8 Ton plus 16 cycles of monotonic horizontal loading. The loads were applied with hydraulic jacks with the skill to apply push and pull actions. The loading was controlled by the lateral displacement at wall top. The 16 cycles (**Figure 4**) were applied in pairs (push and pull actions), from 0.5 to 12 mm; for the specimen SP1, the last two cycles were asymmetric (14 and 27 mm), due to the limitation of the hydraulic jack in performing pull actions. The associated applied loads were above 11 Ton and 9 Ton in push and pull actions. For specimen SP2, the hydraulic cylinder was changed to have the same load capacity in pull and push.

5.2.1 Displacement and rotation of the wall

The vertical transducers TD05, TD03, and TD04, located under the specimen slab, were used to measure the vertical displacements during the test. These transducers were installed along the centerline under the wall (**Figure 2**). Due to the effect of horizontal loading, each time the right transducer (TD05) rises above the original reference line, a rigid body wall rotation occurred (**Figure 3**). We observed a considerable increased in the vertical displacement with every incremental load cycle, indicating that due to the intensity of the vertical load on the slab, the observed overall structural performance was nonlinear. In each cycle, the slab experienced a displacement of 0.5 mm without returning to its initial position. This represented a cumulative displacement of approximately 9.5 mm at the end of test.

5.2.2 Shear strains

The wall rotation, γ, caused by shear deformations in specimen SP1, is established according to the mechanics of materials theory. Because transducers installed in the wall diagonals did not register any deformation during the first stage of the loading process (first six loading cycles), therefore, it can be concluded that the wall did not develop any shear strains in this phase. As a consequence of slab rotation, the wall only experienced base rotation and bending. But, in the seventh cycle, diagonal transducers, TD011 and TD012, began to register shear strains. Still until the tenth cycle, the masonry wall behaved linearly, but later the wall suffered nonlinear deformations.

5.2.3 Global and relative horizontal drifts

The total distortions, R_{tot}, and the relative drift, R_{eff}, of specimen SP1, measured in the combined loading are illustrated in **Figure 7**. The envelopes of the hysteretic cycles are also shown, which are estimated with the bilinear behavior of the curves. The rigidity of SP1 was defined from the slopes of the curves; the stiffness of the complete system was determined to be equal to 11.3 Ton/cm, while the rigidity of the isolated wall was assessed to be equal to 40 Ton/cm (the effect of the rotation as a rigid body was removed). This stiffness is 3.5 times greater than that due to the wall resting on a flexible base (wall placed on the centerline of a two-way slab).

For second specimen, SP2, it can be observed that the slab rotation is practically equal to the wall bending (**Figure 8**). It can be concluded that the wall did not have any shear deformation.

In **Figure 8**, four limits are proposed to define the behavior of the specimen SP2. A behavior without damage (green color), a behavior with moderate damage (yellow color), a behavior with serious damage that can be repaired (orange color) and a behavior with serious damage that cannot be repaired (red color).

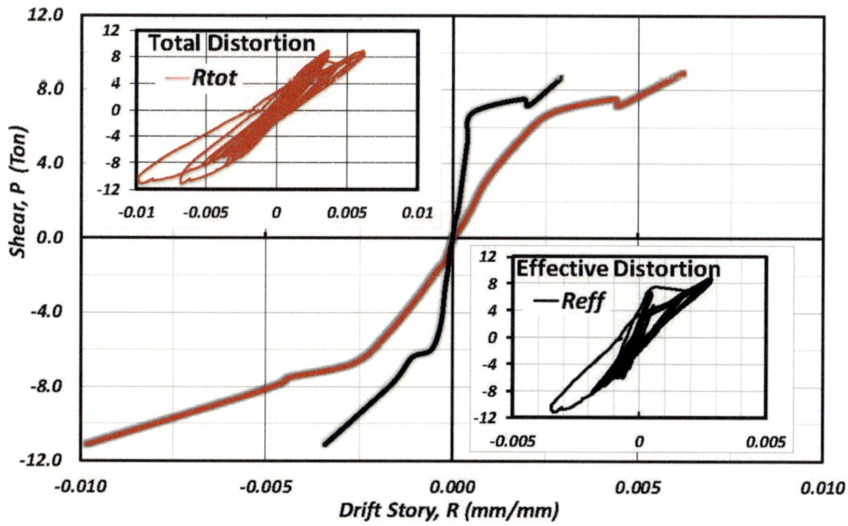

Figure 7.
Global (R_{tot}) and effective (R_{eff}) distortions of specimen 1 (with masonry wall) during the combined load, as well as their respective envelopes.

The experimental stiffness for the second specimen (SP2) was calculated from the slopes of the curves; the stiffness of the slab-wall system was estimated as 10.45 Ton/cm, this value is less than the first specimen (SP1) because the slab had damage prior to the beginning of the experiment.

In **Figure 9** are made two comparations: first, the numerical capacity curve "Effective FEA" (or MBR_M2.5 masonry wall on rigid base) is compared with experimental curve SP1 "Effective EXP"; and second, the numerical capacity curve "Total FEA" (or M10_M2.5_12VL masonry wall on flexible base) with experimental curve SP1 "Total EXP". The comparison between pairs of curves indicates that these are similar at low levels of load, but when comparison is made between rigid and

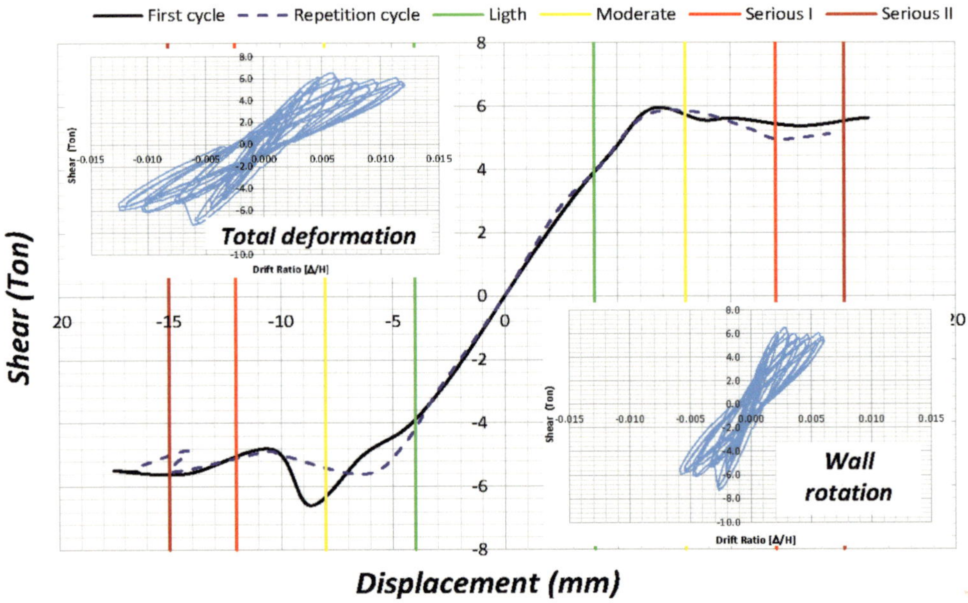

Figure 8.
Distortions of specimen SP2 (RC wall) due to the combined load, as well as their respective envelopes.

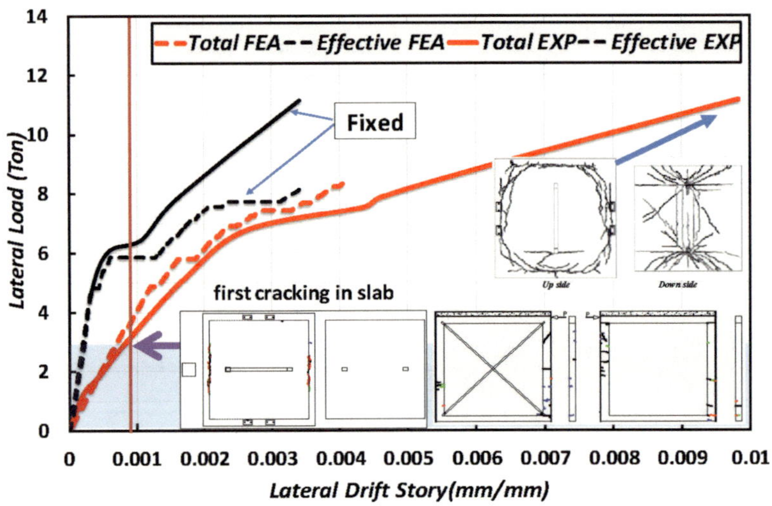

Figure 9.
Comparison between numerical and experimental model SP1. Experimental flexible base (total) and rigid base (effective) versus numerical behavior.

flexible base, we observed significant differences between the slopes of the curves. Then, stiffness results indicated that for a wall on a flexible base a value of 14 Ton/cm was reached, while the one on rigid base was computed as 40 Ton/cm, this represent a ratio of 3 times between both cases. In conclusion, when experimental SP1 curve and the numerical curve were compared for the same support conditions the slopes were similar.

When the capacity curves from RC walls numerical models (**Figure 10**) are analyzed, we observed a great difference in the behavior respect to the one observed in the masonry wall models. The curves of RC walls numerical models M9_C2.5_12VL (flexible base) and MBR_C2.5 (fixed base), show a wide difference in the linear slopes for as many as 15 times; we calculated 14.1 and 210 Ton/cm. Nevertheless, the rigidity values for the flexible models M10_M2.5_12VL (masonry) and M9_C2.5_12VL (RC) are close: 14.1 vs. 11.3 Ton/cm [2, 13]. The high flexibility of the slab does not allow the RC wall to develop its full strength and rigidity. This is an outstanding result that we would like to underline.

5.3 Slab-wall stiffness of specimen SP2

Figure 11a shows peak-to-peak stiffness, which is the slope of the average lateral force vs. total displacement diagram [15]. K stiffness is achieved by joining the maximum displacement points in a cycle along a straight line; the slope of the line is the cycle stiffness, and, is calculated as the difference of the shear forces, divided by the difference of displacements. **Figure 11b** shows the normalized stiffness, calculated as the peak-to-peak stiffness between the first cycle stiffness for each data series, that is, the data series stiffness for the first cycle and for the repetition cycle.

In **Figure 11b** is included the drift story limit, $\gamma = 0.006$, (indicating that nonstructural elements can be damaged). For this distortion limit value, the normalized average stiffness is approximately 35% of the initial moment of inertia, this loss of stiffness is only due to the transfer slab, because the reinforced concrete wall showed no damage. This important result should be considered in the modeling of buildings with transfer slab, so that lateral displacements can be properly calculated.

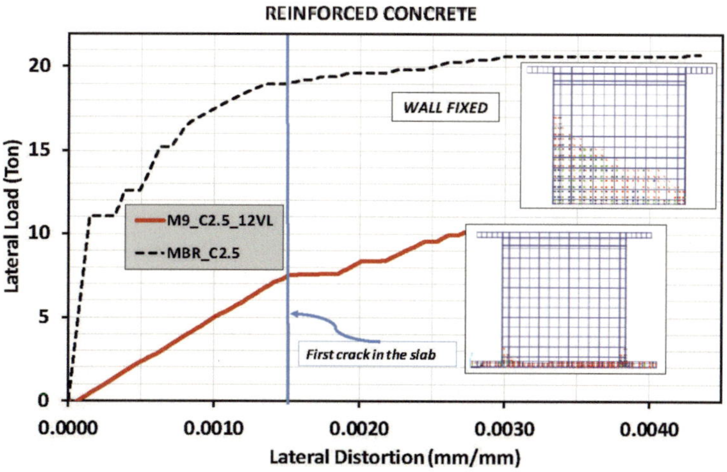

Figure 10.
Load-distortion curves for M9_C2.5_12VL and MBR_C2.5 (fixed) models of RC wall, including crack patterns.

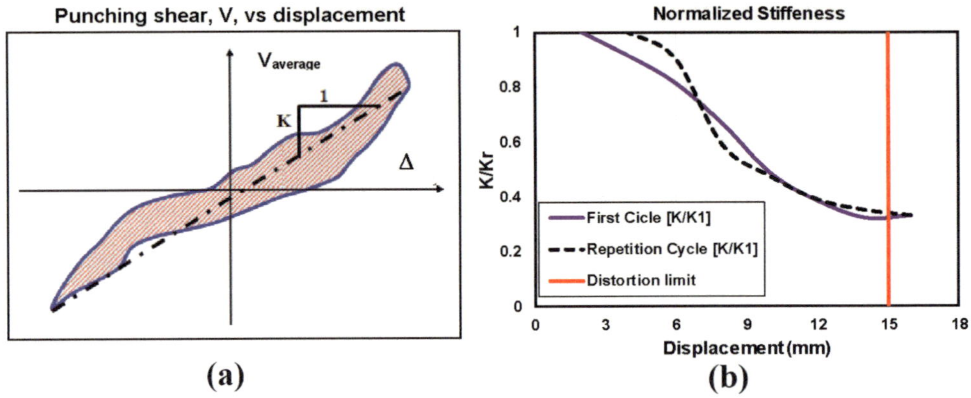

Figure 11.
(a) Peak-to-peak stiffness according to Chopra [15]. (b) Normalized stiffness load-distortion curves for M9_C2.5_12VL and MBR_C2.5 (fixed) models of the concrete wall, including crack patterns.

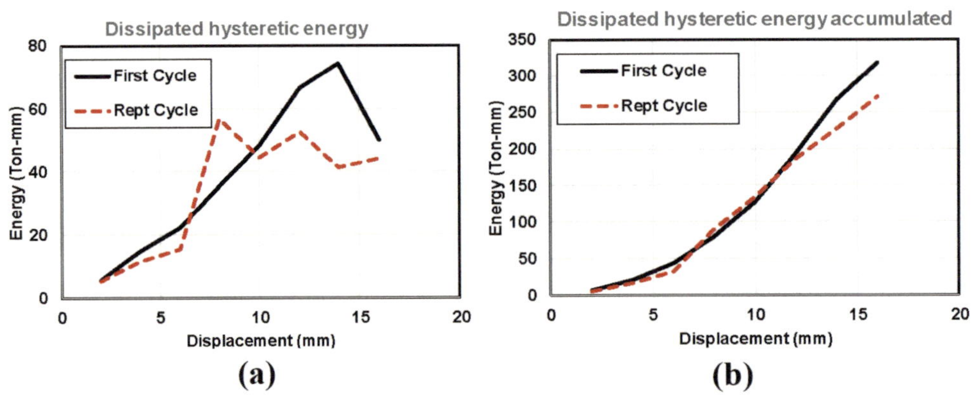

Figure 12.
(a) Hysteretic energy dissipated by cycle and (b) accumulated dissipated hysteretic energy.

5.3.1 Hysteretic energy dissipated in specimen SP2

Figure 12a shows the hysteretic energy dissipated by cycle, for each data series (first cycle and repetition cycle). This energy was calculated for each cycle as the curve enclosed area "Average shear force" vs. " Total displacement" (shaded area in **Figure 11a**). It can be observed that the energy is increased with the lateral displacement. There is a change in the energy slope when the displacement is 14 mm, the energy decreases, probably due to the accumulated damage in the transfer slab. **Figure 12b** shows the accumulated dissipated hysteretic energy for each data series (first cycle and repetition cycle). It can be observed that the curve slope of the first cycle decreases, for the reasons mentioned above.

5.3.2 Equivalent viscous damping

The amount of equivalent viscous damping [15] corresponding to the energy dissipated by the structure through its nonlinear response at every cycle can be determined using the response of Specimen SP2. The equivalent viscous damping can be calculated in experimental curves, with Eq. (5). This parameter represents the internal frictions in the material [15] that, in the case of the studied specimen

SP2, the cracks are concentrated in the transfer slab. Equivalent viscous damping is estimated as:

$$\zeta_{eq} = \frac{1}{4\pi}\frac{E_D}{E_{S_o}} = \frac{1}{4\pi}\frac{\text{Energy dissipated}}{\text{Strain energy}} \tag{5}$$

Where, E_D is the energy dissipated and is calculated by the area enclosed by the hysteresis loop, and E_{So} is the dissipated strain energy.

Equivalent viscous damping is also used as an index that measures the damage accumulated in the member; as the crack increases, so does the damping. Mexican Building Code assumes an equivalent viscous damping equal of 5 percent of the critical, to use in seismic design. **Figure 13** shows that the average damping in the connection is about 6%, which is a value similar that assumed in traditional design.

It is also possible to estimate the ductility, which is determined from the enveloping graph of the hysteresis cycles, i.e. the curve "Lateral force" vs. " Relative displacement". The graph is idealized as an elastoplastic curve, the elastic part is defined as the line joining the origin point of the curve with the lateral load point corresponding to two thirds of the maximum shear recorded in the test (V_{test}); the idealized plastic portion of the curve, envelope the maximum load and ends up to the ultimate displacement, which is defined as the value corresponding to 80% of the maximum resistance recorded curve. Connection ductility (μ) can be determined as the ultimate displacement (Δ_u) divided by the yielding displacement (Δ_y).

The minimum value of ductility in specimen SP2 was calculated equal to 2.19. This result illustrate that high values of ductility reduction factors cannot be adopted in the design of buildings with transfer slabs.

5.4 Analysis of crack patterns

5.4.1 Cracking in the walls

The crack pattern of masonry wall of SP1 is show in **Figure 14**, the cracks had an orientation similar with typical patterns of masonry walls subject to shear loads. This distribution of cracks is typical in confined masonry walls where the cracks are in wall diagonals, and regardless of how the wall is supported, flexible or fixed. Whereas, the crack pattern on the RC wall of specimen SP2 is illustrated in **Figure 14**, it can be observed only a few minor cracks located in the top of the wall, practically without cracks (**Figure 14**), this because the horizontal force is transmitted as a shear

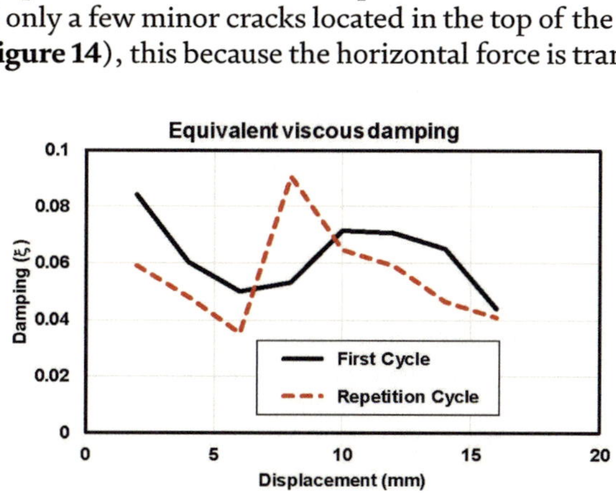

Figure 13.
Equivalent viscous damping in specimen SP2.

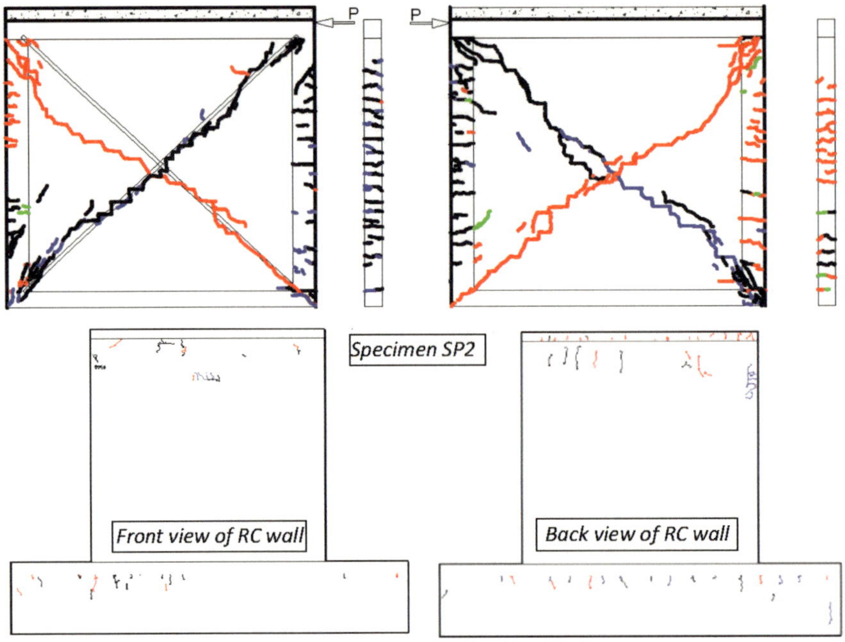

Figure 14.
Cracking on the masonry wall (top) and on RC wall (bottom), after applying the complete load cycles. Cracking on the front side and back side of the wall is illustrated.

in the most resistant direction of the wall and in the slab is passed as a bending with respect to the minor inertia axis.

We also studied the pattern and distribution of cracking in the numerical models. Cracks in the masonry wall of experimental prototype SP1 are similar as the ones observed in the slab-wall numerical models. In the other hand, the behavior of the RC wall of specimen SP2 was according with the expected, it is to say, only minor cracks, the pattern is very similar with the numerical model of **Figure 10**; In the case of RC walls, the crack pattern was quite different when the base condition is analyzed; the wall on a flexible base (transfer slab) only showed a few cracks, while a wall resting on a fixed base, the crack pattern is concentrated in the tension zone of the wall (lower left corner). RC wall on a fixed support demanded its full shear capacity, while the wall on a transfer slab developed only some shear strains. This situation can be attached to the rotations of the slab, which caused a great decrease in the rigidity capacity of the entire system, **Figure 15**; a situation very delicate for the slab, particularly in the zones at the ends of the wall. The slab presented in this zone large cracks on its lower side and crushing of concrete on the upper side. This condition diminishes the slab capacity significantly (**Figure 15b**).

5.4.2 Cracking in the slabs

During the combined loading stage in prototype SP1, first cracks were detected at slab edges, close to the contiguous girders, and perpendicular to the wall. But, when the load was increased, cracks trajectories at both faces of the slab were detected (top of **Figure 15**). Cracks in the perimeter edges of upside slab were delineated, and other cracks were detected at the edges of masonry wall (**Figure 15**, top-left and bottom-left). Nevertheless, in the downside of slab, the cracks were defined as parallel lines to the walls and other lines with radial trajectories, most of these last cracks had a common origin located in the connections

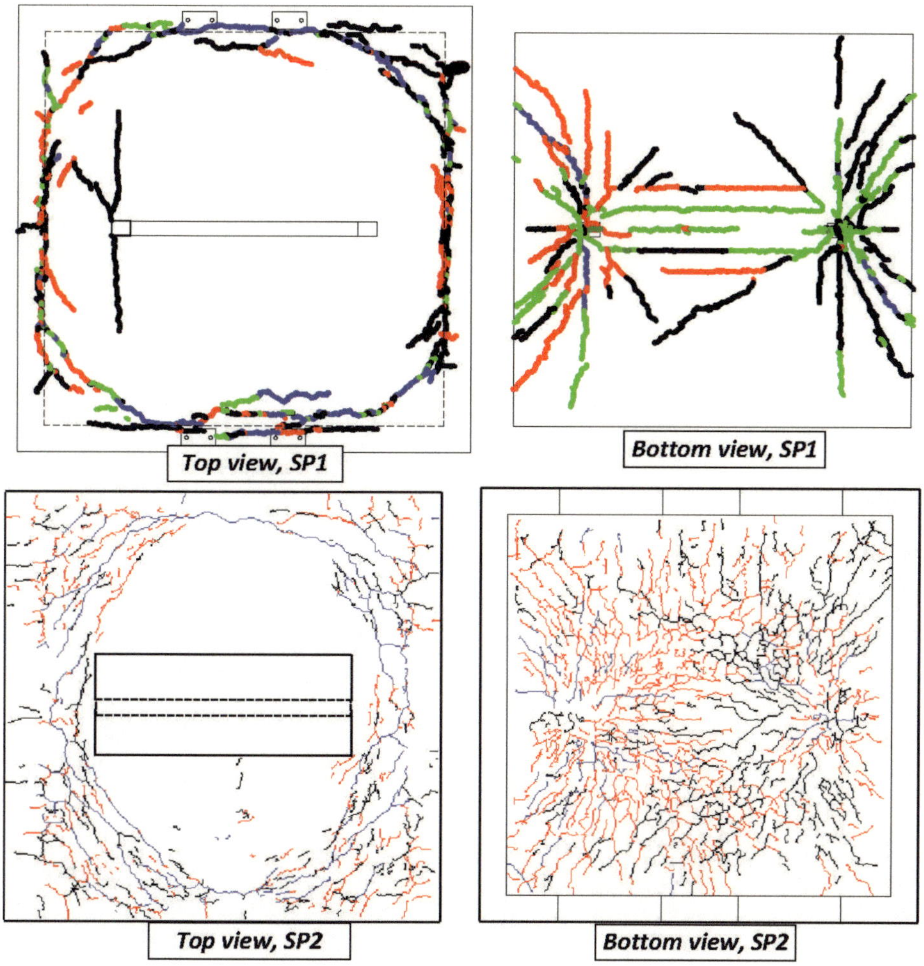

Figure 15.
Cracking in the masonry wall-slab prototype SP1(top) and cracking in the RC slab after applying the complete load cycles in the RC wall-slab prototype SP2 (bottom).

between tie-columns and slab. In the specimen SP2 (bottom of **Figure 15**) was observed a significant damage, in both sides the path of cracking was similar of SP1, however the damaged was more extensive.

5.5 Repaired specimen SP2

Due to the reinforced concrete wall was not damaged, the decision was made to repair the tested Specimen SP2. The repair consisted in the incorporation of two C shaped steel channel (**Figure 16**). The reinforcement system was designed neglecting the load capacity of the slab. Considering a vertical load of 8 Tons applied to the wall and a lateral load applied to the upper end of the wall of 6.5 Tons. The beams were considered fixed at their ends.

In the bending design of steel beams, we assumed a member with a simple section. Stud shear connectors were placed in the steel profiles to achieve a composite section (**Figure 17**). The steel profiles were fixed to the perimeter beams by steel plates that were fixed by fasteners. A space of 1 cm was left between the concrete slab and the steel profile to fill it with grout, thereby ensuring that the concrete slab is in contact with the profile along its entire length. The concrete

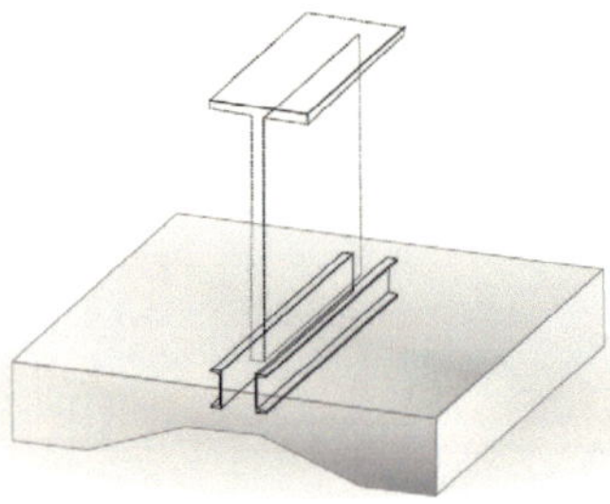

Figure 16.
Reinforcement system with two C-shaped steel channel.

Figure 17.
C-shaped steel channel with stud shear connectors (left), drills in the concrete slab.

slab was drilled so that the stud shear connectors could be placed (**Figure 16**), finally the grout was placed.

When the second specimen, SP2, is repaired, it is possible to increase some of the most important mechanical properties, such as strength, rigidity, and deformability (**Figure 18**). The resistance of the original specimen is 6.59 Tons while the resistance of the repaired specimen is 15.4 Tons, that is, it increased 2.34 times. The initial rigidity of the original specimen is 2.64 Ton/m, while the rigidity of the repaired specimen is 4.32 Ton/m, that is, an increase of 64% was achieved. The maximum displacement in the first specimen was 16.1 mm, which corresponds to a 0.6% drift ratio, the maximum drift ratio in the repaired specimen was 45.15 mm, which corresponds to a drift of 1.8%, that is, the capacity of displacement was tripled.

By neglecting the contribution of the concrete slab, a safety factor is achieved in the specimen that improves the mechanical properties.

It is concluded that the repair with steel profiles placed under the slab, is effective in improving the mechanical characteristics of rigidity, strength, and displacement capacity. This type of reinforcement can be used on transfer slabs that do not have sufficient capacity for the demands to which they can submit and prevent the damage.

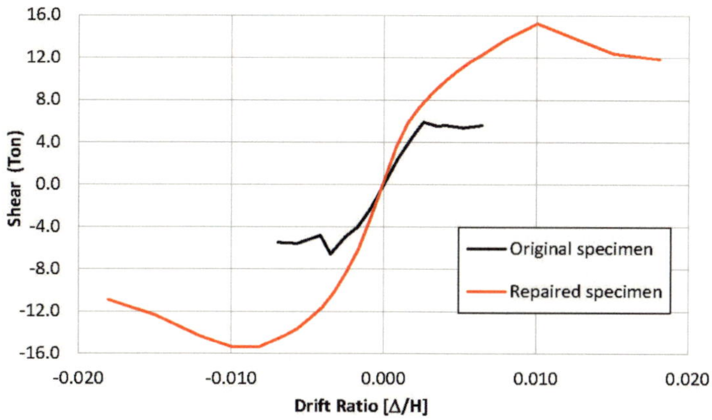

Figure 18.
Comparison of hysteresis envelopes of the original and repaired specimen SP2.

6. Bending analysis

6.1 Measured strains and calculated bending moments

Bending moments in the two-way slab of specimen SP1 subject to the load wall can be found using the nominal moment capacity of a singly reinforced rectangular strip, with a width defined according to NTC [12], and according with the slab reinforcement of SP1 (**Figure 3**).

In **Figure 19** are illustrated the strain variations measured from vertical load stage along the two principal central strips of SP1 slab model (longitudinal and normal wall directions). These strains were measured with the strain gages set installed in the slab (**Figure 3**). The first cracks in the slab (**Figure 9**) appeared when the vertical load reached a value of 4.8 Ton. **Figure 19b** also show the bending moment diagrams associated to the strains along central strips; additionally, the moment diagrams were compared with those computed using Finite Element Analysis. Maximum bending moments are in the joint of slab with tie column zone. In a lateral strip located at 75 cm from the centerline, the measured strains were approximately 60% of those of the central strip, consequently maximum negative moment also had this reduction.

In the horizontal load case, strain variations and bending moment diagrams along the center strips of two directions of the slab are shown in **Figure 20**; these

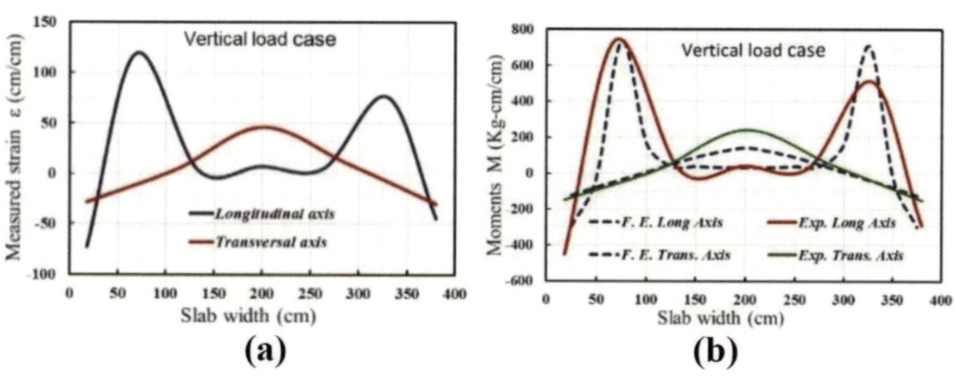

Figure 19.
(a) Strain variations in central strip in longitudinal direction, vertical load case; (b) bending moment in central strip.

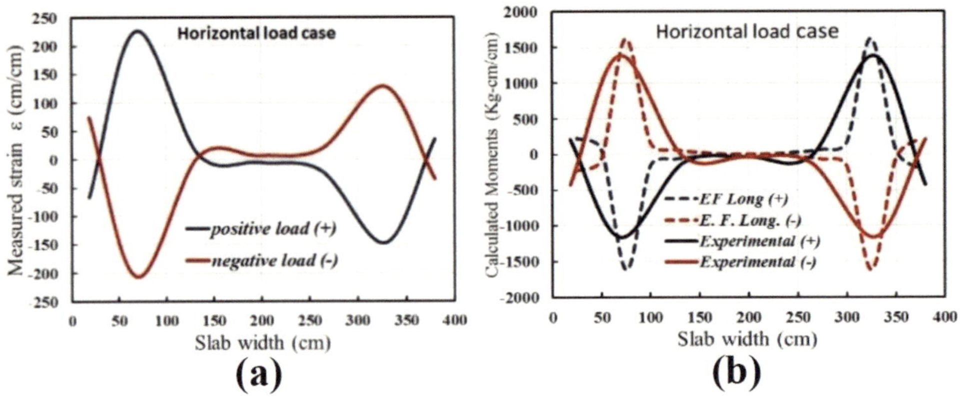

Figure 20.
(a) Strain variations in central strip from horizontal load case; (b) bending moments in central strip.

strains were obtained from the strain gages set, when the horizontal load reached 2.8 Ton. In **Figure 4b**, bending moment diagrams are compared with those computed with a finite element result. It is worth noting that, for this load level, RC slab of SP1 reached the nominal moment strength in central strip. Since maximum positive and negative bending moments are located always in the tie column zone, special design is essential in this zone when transfer slabs are used in structures.

7. Case study: a building constructed with a transfer floor system

Construction of buildings structured with transfer floor systems has become a common practice in many cities in Mexico [16] and other countries [17], some of which have a high seismic risk. To study the discontinuity effect in walls due to transfer slabs in buildings, this section presents some results from structural analysis of one real building constructed in Mexico City for residential use.

Building A has six stories on the ground level and a basement walls as part of mat Foundation. The ground story (top of **Figure 21**) is intended for parking, and their structure is based on RC frames; the five remaining stories are structured with load-bearing walls some of red brick masonry and others of reinforced concrete. Floor slabs are of various types, so the Mat foundation plate is reinforced concrete with concrete walls, the first floor (transfer slab) also is reinforced concrete cast on site, but the remaining upper levels were constructed with prefabricated beam and block, lightened with polystyrene. The plan is shown in **Figure 21**, the area enclosed by the axes A–K and 8–12, is the part of the building which rests on the transfer floor system, which is formed by beams and a solid slab of 12 cm. Most of the interior walls are interrupted at that level and do not rest on any beam (*black line* in figure).

In basis on structural analysis from the real building, shear force ratios in the masonry walls (ratio between ultimate shear and strength shear, V_u/V_D) is presented in **Figure 21**. Graphs in both directions of the building show that resistance is exceeded on a lot of discontinuous walls, and in many walls, shear force ratios are greater than 2.

8. Discussion

The main goal of the experimental program executed was to specify the behavior of wall-slab models for cyclic vertical and lateral load especially it was

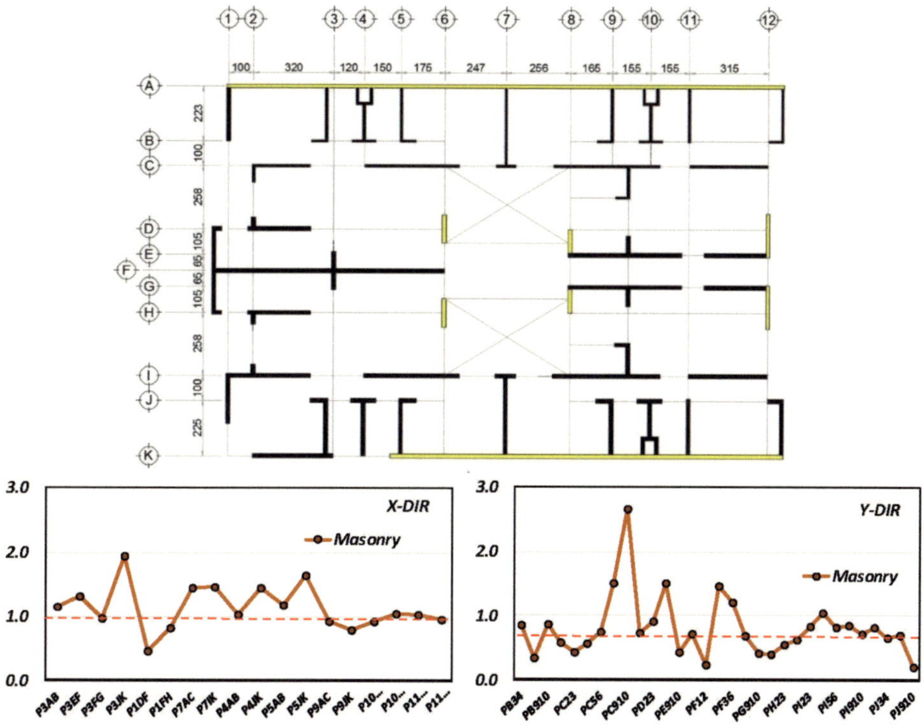

Figure 21.
Transfer floor plan in building A, and shear ratio (V_u/V_D) in walls located on the transfer floor. In black line are indicated the interrupted walls in transfer slab.

demonstrated that using discontinuous walls significantly reduces the strength and rigidity of walls supported by transfer floor, which could be a design limit in practical earthquake engineering.

We propose that transfer slabs can generate the horizontal irregularity out of plane offset that can increase in a significant way the building vulnerability, this condition is different of other structural irregularities as soft story and weak story.

The most relevant finding from real buildings analysis, is that when wall discontinuity is increased, i.e., as there are more walls that are interrupted and supported on the transfer floor system, the shear and axial forces in load-bearing walls are increased too, reaching values that exceed the nominal resistance and design up to more than 3 times. Then it is recommended in the conceptual design of buildings, that the total area of discontinuous walls never can exceed 40% in a main direction; neither more than 30% in the two directions simultaneously.

Further experimental studies are needed that include other types of floor systems used in buildings such as the waffle slabs.

9. Conclusions

The findings of the experimental-analytical research program resulted in the following conclusions:

1. In the first stage, we observed the first cracking in the slab of SP1 when the vertical load reached 5.8 Tons, with a recorded deflection of 0.5 mm at the slab centerline; this load was very close to the design load from the Mexican Building Code. After the first stage, the SP1 regained its elastic behavior.

The numerical model of ANSYS also predicted the onset of cracking at the same load value.

2. Lateral stiffness of SP1 was a third part (33%) of the one observed for a wall on a fixed base. The model had horizontal displacement mainly due to slab bending (66%) and wall bending (33%). In the numerical model, the RC wall stiffness supported on a slab decreases by about 15 times due to the slab flexibility.

3. In the case of SP2, a similar loss of stiffness is observed; when the distortion limit is reached, the normalized average stiffness is approximately 35% of the initial moment of inertia.

4. We noted that the RC wall of SP2 did not show visible damage, unlike masonry wall of SP1, but in both cases, the damage to the slabs was such that these members reached a cracking condition on both slab faces. In SP1, the slab suffered a residual deformation in their center line and developed a membrane action.

5. When we analyze the results of both experimental tests, we concluded that solid slabs with small depths should not be used to transfer shear wall loads, whether masonry or CR, to nearby beams or to columns.

6. The amount of energy dissipated by the specimens was determined with the equivalent viscous damping. In SP2, the cracks were concentrated in the transfer slab. The average damping in the connection was a little more than 5% indicating that these systems dissipate a great amount of energy.

7. When are compared the curves from the experimental and numerical models, we observed an excellent correlation between these two curves.

Author details

Alonso Gómez-Bernal*, Eduardo Arellano Méndez, Luis Ángel Quiroz-Guzmán, Hugón Juárez-García and Oscar González Cuevas
Universidad Autónoma Metropolitana, Azcapotzalco, México City, Mexico

*Address all correspondence to: agb@azc.uam.mx

References

[1] Gómez Bernal A, Juárez García H, Roldán Islas J. Earthquake damage due to soft soil conditions and structural features of the building infrastructure in La Colonia Roma, Mexico City. In: 6th Structural Engineers World Congress 6SEWC. Cancún, QR, México: Structural Engineering Worldwide; 2017

[2] Gomez-Bernal A, Manzanares DA, Juarez-Garcia H. Interaction between shear walls and transfer-slabs, subjected to lateral and vertical loading. In: Proc. Vienna Congress on Recent Advances in Earthquake Engineering and Struc. Dynamics Pap 447, Vienna, Aut. 2013. pp. 28-30

[3] Roeslin S, Elwood KJ, Juarez-Garcia H, Gomez-Bernal A, Dhakal RP. The September 19th, 2017 Puebla, Mexico earthquake. In: Proceedings of the New Zealand Society for Earthquake Engineering Annual Technical Conference. Auckland, New Zealand. 2018. Available from: https://researchspace.auckland.ac.nz/handle/2292/44962

[4] Carrillo J, Alcocer S, González G. Deformation analysis of concrete walls under shaking table excitations. J. DYNA. 2012;**79**(174):145-155. ISSN: 0012-7353

[5] Gouveia JP, Lourenço PB. Masonry shear walls subjected to cyclic loading: Influence of confinement and horizontal reinforcement. In: Proc. Tenth North American Masonry Conference, St. Louis, USA. 2007

[6] Preti M, Migliorati L, Giuriani E. Experimental testing of engineered masonry infill walls for post-earthquake structural damage control. Bulletin of Earthquake Engineering. 2015;**13**(7):2029-2049. DOI: 10.1007/s10518-014-9701-2

[7] Haris I, Farkas G. Experimental results on masonry infilled RC frames for monotonic increasing and cyclic lateral load. Periodica Polytechnica Civil Engineering. 2018;**62**(3):1-11. DOI: 10.3311/PPci.10715

[8] Meli R. Mampostería estructural, la práctica, la investigación y el comportamiento sísmico observado en México. CENAPRED reporte Seguridad sísmica de la vivienda económica, n.17 julio. México; 1994

[9] Park R, Gamble WL. Concrete Slabs. 2nd ed. New York, USA: Wiley & Sons Inc; 2000

[10] Vecchio F, Tang K. Membrane action in reinforced concrete slab. Canadian Journal of Civil Engineering. 1990;**17**:686-697

[11] Xiao Y, Li B, Fujikake K. Experimental study of reinforced concrete slabs under different loading rates. Structural Journal. 2016;**113**(1):157-168. DOI: 10.14359/51688067

[12] Reglamento de Construcciones del Distrito Federal. NTC de Mampostería. Gaceta Oficial del Gobierno CDMX, México; 2017

[13] Gómez-Bernal A, Manzanares D, Vargas O, Arellano-Méndez E, Juárez-García H, González-Cuevas O. Experimental behavior of a masonry wall supported on a RC two-way slab. J. DYNA. 2015;**82**(194):96-103. DOI: 10.15446/dyna.v82n194.46333

[14] Tickoo S, Singh V. ANSYS 11.0 for Designers, CADCIM Technologies, 2009

[15] Chopra AK. Dynamics of Structures. 4th ed. Pearson; 2012

[16] Gómez Bernal A, Juárez García H, Alcántara P, Roldán Islas J. Vulnerability

of buildings with out of plane offsets irregularity constructed in México. In: Proc. XVI World Conf. on Earthquake Engineering, WCEE, Santiago Chile. 2017

[17] Lande PS, Takale P. Analysis of high-rise building with transfer floor. International Research Journal of Engineering and Technology. 2018;**05**(5):2483-2488. e-ISSN: 2395-0056

Chapter 6

Computational Workflow for Three-Dimension Printing in Construction: Digital Tools and Methodological Limitations

Anqi Shi, Sara Shirowzhan and Samad M.E. Sepasgozar

Abstract

Three-dimensional printing in construction (3DPiC) is known as a trending technology in the construction industry. While scholars and practitioners seek to learn more about the applications of 3DPiC, there are no efficient workflows and open data sets available for further investigations. This paper intends to present the data produced in a laboratory for creating new models. The paper first presents the experimentation data collected from 60 models, and selected thermal digital images can be used for further sustainability analysis. The recorded data includes the time of crafting each layer of the model, the total time of creating a model and thermal measures. Based on the 60 experimentations and an intensive literature review, the paper presents a proposed computational workflow, including the use of Revit, Dynamo, Fusion 360, Navisworks and a selected 3D printer, which can be utilised for further data collection and analysis in the field. This model will assist in automating the cost estimation as an upgrade for 3DPiC. This paper is helpful for scholars and practitioners since it shows how laboratory data can be helpful for construction operation design.

Keywords: 3D printer (3DP), 3DP in construction (3DPiC), digital model, computational workflow, clay, material, building information modelling (BIM), Revit, dynamo, fusion 360, Navisworks, quantity surveying, leanness, waste

1. Introduction

3D printing (3DP), namely additive manufacture, was introduced many years ago. The use of 3DP applications has risen recently in various disciplines [1–4], but still is very low in the construction industry due to many challenges and limitations. Furet et al. [5] state that the application of digital technologies including building information modelling [6, 7], virtual reality [8], laser scanner and lidar ([9–11], Sepasgozar et al., [12–16]) is increasing in construction [17, 18]. Then, Furet et al. [5] suggest that the next step will be the implementation of 'three-dimensional printing in construction' (3DPiC) on a wide scale.

The utilisation of 3DPiC can be highly beneficial to the construction industry in several ways, including the possibility of topologic optimisation offering

IntechOpen

customised forms fit purpose, the possibility of complex geometric parts, thermal optimisation and 'leanness' such as reduction of labour and material wastes [5]. However, the process of technology implementation in construction is challenging and needs to be investigated by conducting many empirical experiments [19]. The process of technology adoption is generally complicated and can be affected by many factors, including usefulness, ease of use, vendors support and organisation policy and the users' current infrastructures [15, 18, 20–22]. This paper aims to explore the challenges of 3DPiC by creating several laboratory models. Niemeläa et al. [23] present the result of some of the created models to show how curvature and unique designs may affect the construction operation process where complex curves and forms should be created on sites without using formworks or other traditional tools. This paper also intends to present a proposed workflow for computing the cost of the model in real time. This will be further examined in future investigations.

Figure 1 shows different digital tools, operation methods and materials of 3DPiC. There are six different operation methods, for instance, D-shape and contour crafting (CC) and concrete printing. D-shape was developed by Enrico in 2007 [24], and materials used by this technology are sand, salt and an inorganic binding agent [1]. The CC was introduced by Khoshnevis in 1998 [25] and uses a computer-controlled robot or crane to work efficiently with a short setting time and optimised hydraulic ratio [1]. Concrete printing is a useful method for concrete mortar [1]. This paper adopts the CC method to produce small-sized models in laboratory. This paper presents the research method flowchart, recorded observations during the experimentation process and the proposed computational workflow for future evaluations.

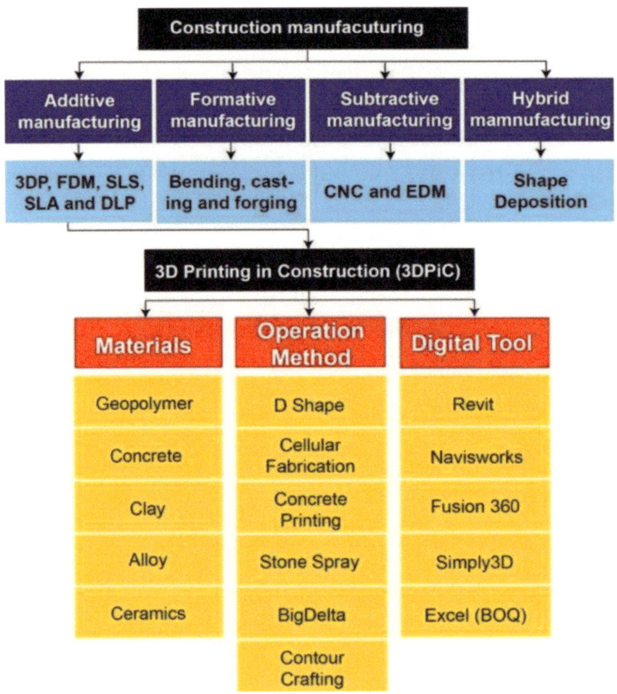

Figure 1.
Digital tools, operation methods and materials for 'three-dimensional printing in construction' (3DPiC). Note: FDM: Fused deposition modelling; SLS: Selective laser sintering; SLA: Stereolithography; DLP: Digital light processing; CNC: Computerised numerical control; EDM: Electrical discharge machining; BOQ: Bill of quantities.

This chapter aims to present a workflow that is developed based on the investigators' experimentations with the support of a wide range of data and records, including 60 laboratory-scale models. These models are structural elements and architecture models in different sizes by using different clay materials. The proposed computational workflow considers Revit, Navisworks and Fusion 360 for estimating the differences between the virtual and physical models and can be useful for estimating the cost of the 3DP models.

2. Research method and data collection

In order to explore the challenges of creating complex geometrical forms and digital tool functions, a series of experiments are conducted. The Mix method, which focused on both qualitative data and quantitative data, was adopted for this research [26, 27]. This paper adopted the quantitative research method as the aim of the laboratory experience is to turn the laboratory observations into statistical analysis. Qualitative methods were chosen due to scarce research studies in this area, and it is essential to introduce a new set of data of the 3DPiC [26, 27]. In order to record the observations, two pieces of equipment and three software programs were used and will be briefly presented in the following section. The data description and analysis are presented in the following subsections.

2.1 Tools and equipment

A 3D printer, namely Potterbot SLX-2 (Scara) printer (see **Figure 2 (a)**), was used to create 60 models, and a thermal scanner 875-2i (see **Figure 2 (b)**) was used to measure thermal factors such as temperature and carbon.

2.2 Data description

The data is collected through printing 60 laboratory-scale models, comprising 56 models created by clay and four created by 'paper clay'. There are 32 of them that are structural elements such as arches, rectangles and squares in different scales.

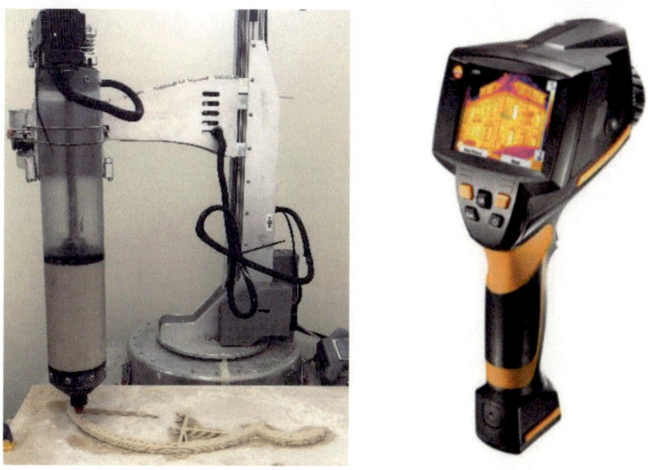

(a) (b)

Figure 2.
Tools used in this research: (a) Potterbot SLX-2 (Scara) printer; and (b) thermal imaging scanner 875-2i.

The four paper clay models are the architecture models. Moreover, 14 models of them are the columns in different twisting degree and the cylinder in different slope degree.

Two models were printed twice due to the 3D printer running out of clay during the printing process. The models could not be continuously printed after refilling the extruder with clay because the previous layers of the model had hardened quickly. The thermal digital images were collected five days after producing each model by Testo 875-2i for the thermal and humidity analysis also can be used for further sustainability analysis. These images can be used for further analysis, such as building material analysis, human comfort analysis, insulation issues, connection issues between the layers, the energy embodied and radiation, design optimisation analysis, scheduling progress management of printing and post-maintenance.

2.3 Data collection

Figure 3 shows the selected images of nine different models created in the laboratory. There were waste materials produced during the experimentations due to the nozzle travelling path, for example, **Figure 3 (c–h)**. This shows that the optimisation of the path and experimentations on smaller scales are required before the full-sized practice for each written coding to ensure the arm and nozzle path will be desirable with less waste materials.

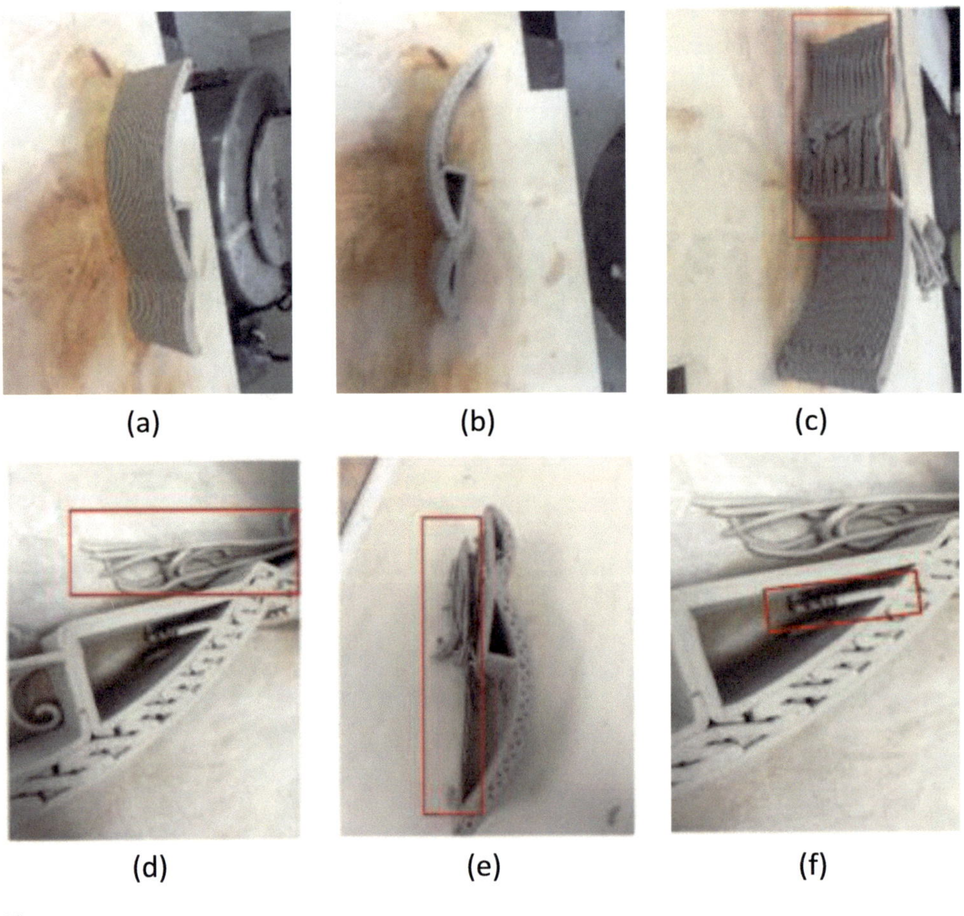

Figure 3.
Arches models at different scales. (a–c) Arch 50% of original size; (d–f) arch 80% of original size.

The flaws in the models are caused by bubbles errors. There were some air bubbles in the 3D printer extruder, and when the materials that involve the air bubbles had been push out, the small flaws and sound were produced. In addition, producing complicated forms was also challenging, and the designed shapes were not produced, as shown in **Figure 4 (a–b, d–g, i** and **k–l)**. The red rectangles shown in the photos refer to waste materials produced during the casting process. The observations show that information flow and controlling the machine to produce desirable forms require a more efficient workflow to save the materials and also for cost estimation.

There are thermal digital images of one of the paper clay architecture models (see **Figure** 5), and different colours show different temperature ranges. The humidity digital images are attached in the appendix.

Table 1 shows the recorded time per layer for 14 models, with columns in different twisting degrees and the cylinder in different slope degrees, and it does not

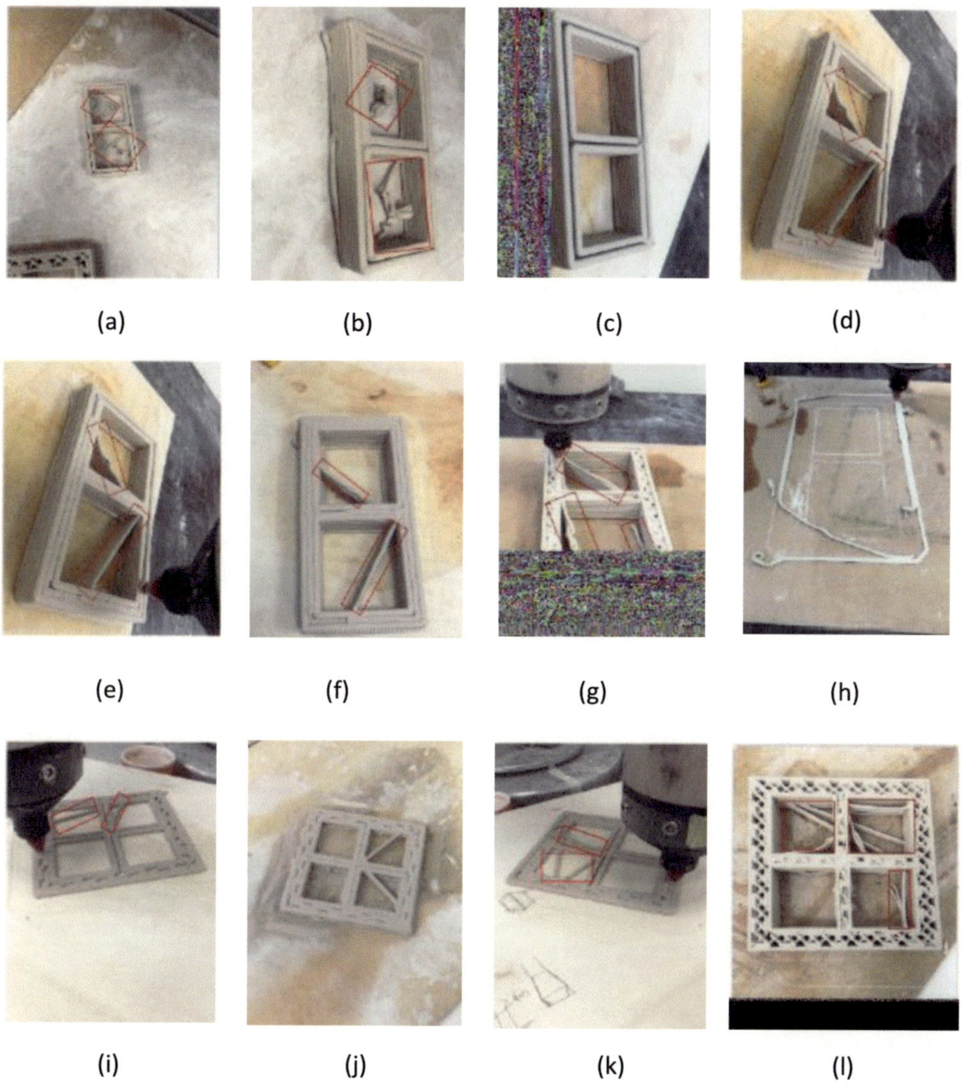

Figure 4.
Rectangle-shaped models at different scales. (a) Rectangle 100% of the original size; (b) rectangle 110%; (c) rectangle 120%; (d-e) rectangle 130%; (f) rectangle 140%; (g) rectangle 170%; (h) rectangle 180%; (i–l) square 100% of the original size.

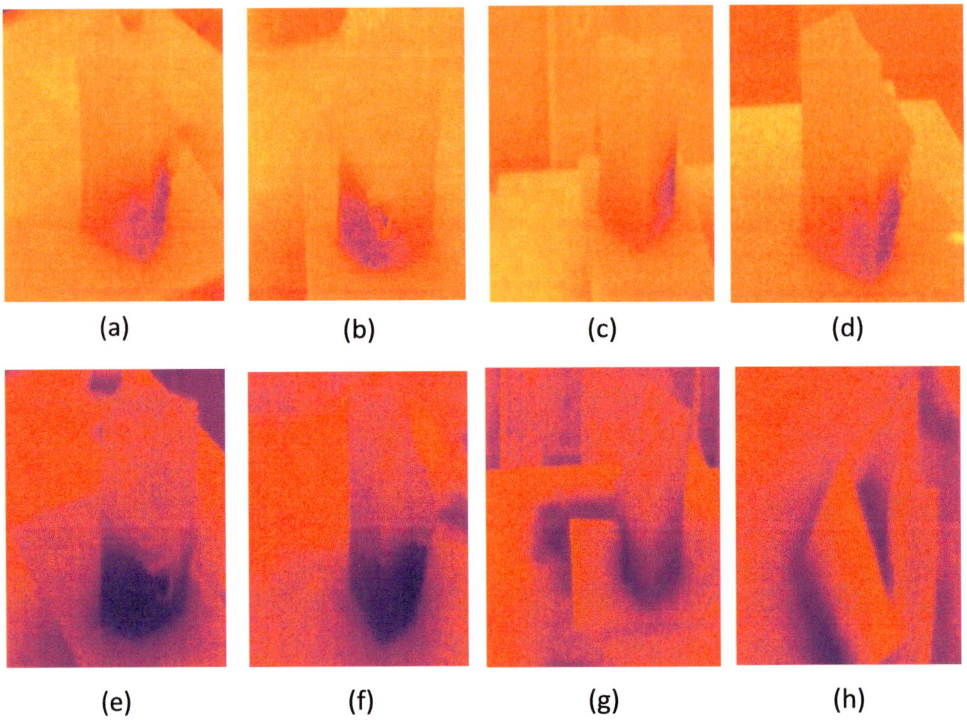

Figure 5.
Thermal digital images of one of the architecture models. (a–d) Thermal image 24–28 degree; (e–h) thermal image 25–29 degree.

Shape	Shape ID	Sample	Scale%	Slope (degree)	Layers	Total Time	Time/Layer
Cylinder	1	1	100	0	25	0:10:29.430	0:00:12.736
Cylinder	1	2	100	10	25	0:08:46.310	0:00:12.110
Cylinder	1	3	100	20	25	0:08:46.650	0:00:11.400
Cylinder	1	4	100	30	22	0:08:31.870	0:00:10.452
Cylinder	1	5	100	45	22	0:07:04.340	0:00:09.155
Cylinder	1	6	100	60	21	0:06:13.560	0:00:06.847
Column	2	1	100	0	60	0:19:12.200	0:00:15.735
Column	2	2	100	10	60	0:18:57.510	0:00:15.752
Column	2	3	100	20	60	0:21:25.450	0:00:15.855
Column	2	4	100	30	56	0:17:58.850	0:00:15.702
Column	2	5	100	45	54	0:19:18.860	0:00:16.000
Column	2	6	100	60/30	60	0:21:10.760	0:00:15.859
Column	2	7	100	70/20	60	0:20:50.800	0:00:15.770
Column	2	8	100	80/10	60	0:21:06.390	0:00:15.720

Table 1.
Characteristics of two models with different slope degrees (14 models).

include the start, priming edge and finish time. The priming edge is used to make the extruder ready for printing, and the finish time is the end running process for the extruder. The design diameter of the cylinder is 10 cm; however, the diameter

Shape	Shape ID	Sample	Scale%	Slope	Layers	Total Time	Time/Layer
Column	1	1	100%	0	100	0:29:14.010	0:00:14.883
Column radian	2	2	100%	0	100	0:30:49.180	0:00:15.874
Column slope	3	3	100%	0	123	0:30:41.840	0:00:12.481
Column dome	4	4	100%	0	97	0:24:35.840	0:00:13.307

Table 2.
Architecture models (recorded time).

Shape	Temperature (degree)	Humidity (rh)	Co2 (ppm)
Column	27,1	75,1%	481
Column radian	27,0	76,1%	456
Column slope	27,0	76,8%	469
Column dome	27,1	76,9%	442
Room temperature	27,1	76,1%	476

Table 3.
*The detailed information for the architecture model in **Table 2**.*

of the actual printed model with a 0-degree wall slope is 9.6 cm. The model with a 20-degree slope had a collapse trend. The most serious deformations occurred at the 30-degree and 45-degree columns.

Tables 2 and **3** show the detail recorded time and the temperature and humidity information for architecture models, respectively. Also, recorded time does not include the start, priming edge and finish time.

3. Proposed computational workflow and future directions

Our experimentations show that controlling the 3DP during construction is challenging and may affect quality due to the unexpected flows and wastage. According to the laboratory observation, most of the 3DP models presented in this chapter have the waste part. Thus, it is necessary to develop efficient workflows to ensure the 3DP process is efficient and easy to estimate the exact volume materials used for 3DP models and further manage the costs. While different algorithms are presented in the literature, including the travel salesman problem algorithm (TSP) used to select the shortest path [28], there are not enough workflows offering an efficient method for controlling and operating the system for construction purposes, especially for materials estimation and costs management. There is an urgent need to develop new workflows that are interoperable with new hardware and software programs. This may increase the operational control of casting processes, the quality of the produced models and calculate the wasted materials. **Figure 6** shows a proposed workflow for the integration of additive manufacture with 5D BIM to compare the cost difference between digital measurement and real-time production. This workflow intends to generate the cost of the produced work with real-time measurement of produced layers and then compares with costs generated based on the measurement from digital software.

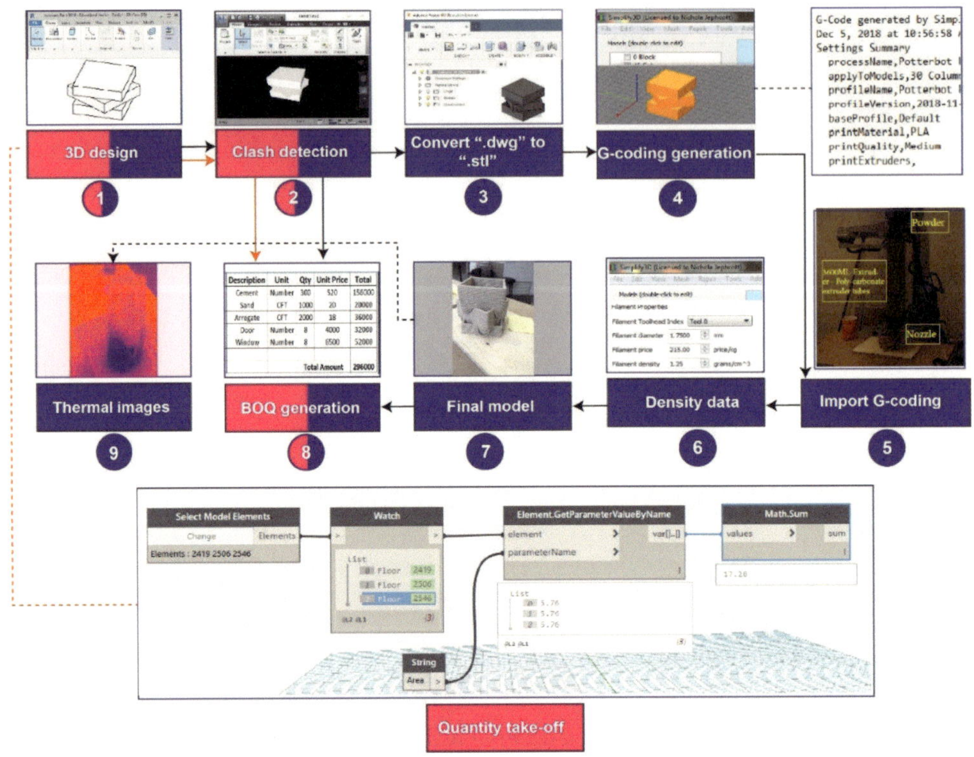

Figure 6.
The proposed workflow for real-time and design digital cost estimation of contour craft objects.

The real-time measurement consists of nine stages:

Stage 1: the 3D model will be created in the Revit; it can be imported into Navisworks by.*rvt* or.*rfa* extension.
Stage 2: clash detection can be done through Navisworks; the schedule and other cost information are collected and imported into the Bill of Quantity (BOQ) [29]. The 3D model is saved as a.*dwg* format because it can be used in Fusion 360. In addition, the design quantity information can be extracted from Revit with Dynamo Extension [30] and then can be imported into BOQ.
Stage 3: import the 3D model into Fusion 360, and convert the model field into.*stl* extension. Then, the CAD model can be used in the slicer software 'Simplify 3D' to generate the G-coding. This is required because the.*dwg* format cannot be used directly for generating G.
Stage 4: Simply3D is used to slice the 3D models into the printable layer; the G-coding automatically generates the required coding.
Stage 5: USB with G-coding into the drive board of a 3D printer.
Stage 6: density (p) is computed by Simply3D, and by measuring mass (m) of the real physical model to estimate the volume (v) in each layer.
Stage 7: final physical model in the desired scale is created.
Stage 8: import the real physical volumetric data to the BOQ, and estimate the actual cost.

A sensor or a material flowmeter can be used during the printing process. There are several different sensors available in practice, such as the concrete flowmeter offered by 'Jiangsu Sipai Instrument Co.' in 2014. The detection range of SP-LDE concrete flowmeter is from 0.0636 to 4521.6 m3/h [31]. During casting, the volumes of casted parts will be measured by the installed sensor, and the data can be compared with the estimation given by Revit.

Stage 9: the thermal images are generated using a thermal scanner. The thermal images used in this paper are recorded five days after the casting process of each model.

Some of the stages can be merged or changed if other hardware or software programs are used. For example, if Dynamo is used, the design quantity information would not be required to extract from Revit. Future studies should focus on examining the proposed workflow and providing more details of each stage. Other new technologies also should be investigated to reveal what type of software or hardware technologies are interoperable and can be used efficiently for 3DPiC.

4. Conclusions

This paper aimed to present a series of data collected during the process of designing and creating 3DP models in different sizes by using different clay materials. In addition, the paper proposed a workflow for collecting required digital data for estimating the physical progress of casting the model, volumes and associated costs.

This is a step forward to fill the gaps in knowledge to control 3D printing in an efficient way, acquire digital data for cost measurements in real time. The workflow can be examined as future studies and also can be revised by combining the proposed software programs with other new technologies to produce more information for progress monitoring and estimating. While the major limitation is that some of the full-size practices cannot be experienced, the flow of digital information can be tested in the laboratory. In addition, the observations and the data presented in this paper recommend that construction practitioners should test the programming codes, and the parametric design at the laboratory scale before the full-scale practice for optimising the movements of the 3D printer arm, quality and waste controls is undertaken.

Acknowledgements

The authors acknowledge the support of the Design Futures Lab of the Faculty of Built Environment at the University of New South Wales, Sydney, Australia, to support this study and provide required materials and the selected robots to produce models.

Dataset: https://doi.org/10.6084/m9.figshare.8427023.v1

Appendix

See **Figure** 7.

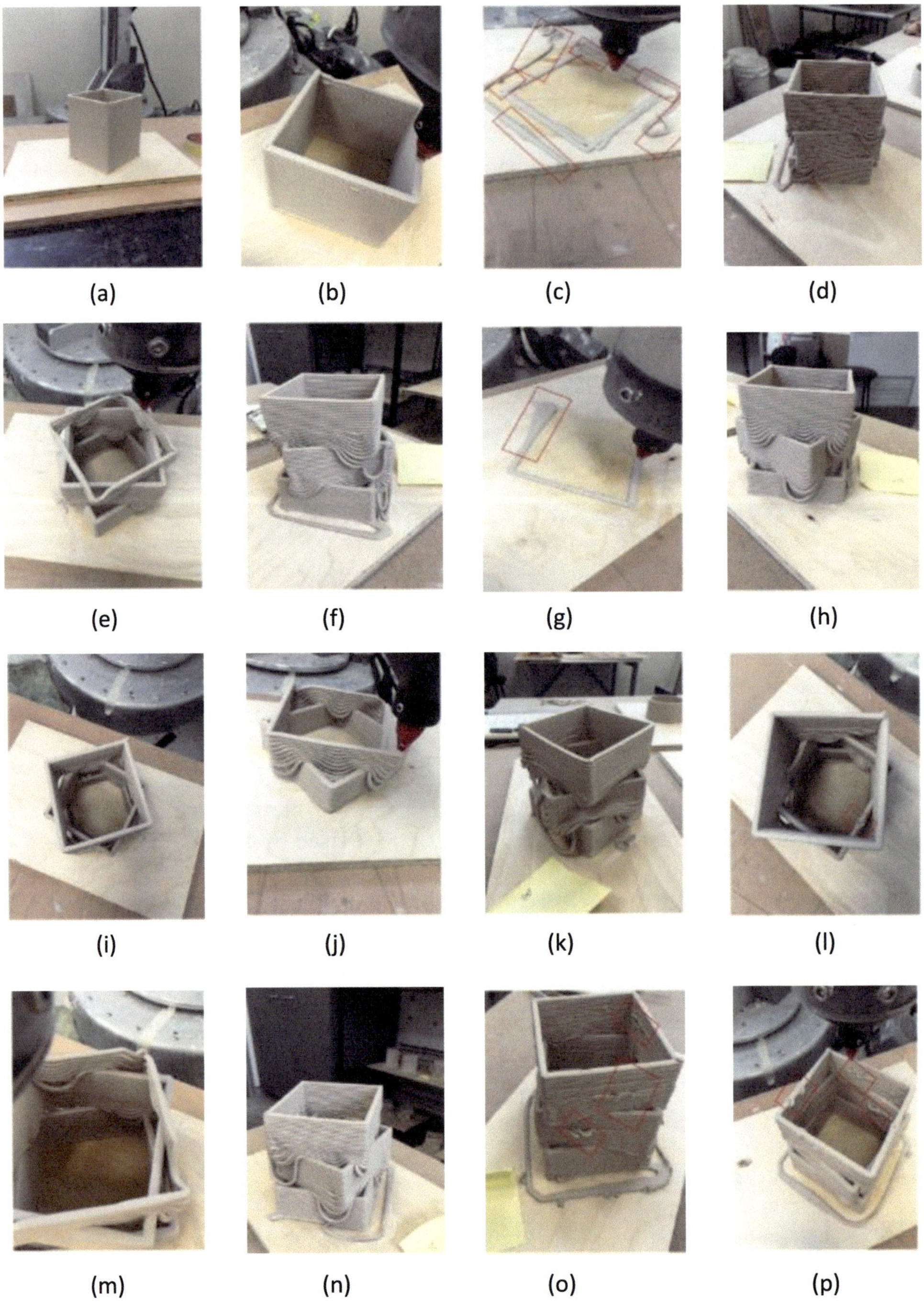

Figure 7.
Column models at different twist degrees. (a-b) Column 0 twist degree of original design; (c-d) column 10 twist degree of original design; (e-f) column 20 twist degree of original design; (g-h) column 30 twist degree of original design; (i-j) column 45 twist degree of original design; (k-l) column 60 (-30) twist degree of original design; (m-n) column 70 (-20) twist degree of original design; (o-p) column 80 (-10) twist degree of original design.

Author details

Anqi Shi*, Sara Shirowzhan and Samad M.E. Sepasgozar
Faculty of Built Environment, University of New South Wales, Sydney, Australia

*Address all correspondence to: anqi.shi@unsw.edu.au

References

[1] Ma GW, Wang L, Ju Y. State-of-the-art of 3D printing technology of cementitious material—An emerging technique for construction. Science China Technological Sciences. 2018; **61**:475-495

[2] Shakor P, Nejadi S, Paul G, Sanjayan J. Dimensional accuracy, flowability, wettability, and porosity in inkjet 3DP for gypsum and cement mortar materials. Automation in Construction. 2020;**110**:102964

[3] Tahmasebinia F, Niemelä M, Ebrahimzadeh Sepasgozar S, Lai T, Su W, Reddy K, et al. Three-dimensional printing using recycled high-density polyethylene: Technological challenges and future directions for construction. Buildings. 2018;**8**:165

[4] Tahmasebinia F, Sepasgozar SM, Shirowzhan S, Niemela M, Tripp A, Nagabhyrava S, et al. Criteria development for sustainable construction manufacturing in Construction Industry 4.0. Construction Innovation. Construction Innovation. 2020;**20**(3):379-400

[5] Furet B, Poullain P, Garnier S. 3D printing for construction based on a complex wall of polymer-foam and concrete. Additive Manufacturing. 2019;**28**:58-64

[6] Hong Y, Sepasgozar SME, Ahmadian AFF, Akbarnezhad A. Factors Influencing BIM Adoption in Small and Medium Sized Construction Organizations. Auburn, USA: 2016 Proceedings of the 33rd ISARC. Proceedings of the International Symposium on Automation and Robotics in Construction. IAARC Publications; 2016. Vol. 33. p. 1, pp. 452-461

[7] Shirowzhan S, Sepasgozar SME, Edwards DJ, Li H, Wang C. BIM compatibility and its differentiation with interoperability challenges as an innovation factor. Automation in Construction. 2020;**112**:103086

[8] Li X, Yi W, Chi HL, Wang X, Chan AP. A critical review of virtual and augmented reality (VR/AR) applications in construction safety. Automation in Construction. 2018; **86**:150-162

[9] Sepasgozar S, Shirowzhan S, Wang CC. A scanner technology acceptance model for construction projects. Procedia Engineering. 2017;**180**:1237-1246

[10] Shirowzhan S, Sepasgozar S, Liu C. Monitoring physical progress of indoor buildings using mobile and terrestrial point clouds. Construction Research Congress. 2018;**2018**

[11] Shirowzhan S, Sepasgozar SME, Li H, Trinder J. Spatial compactness metrics and Constrained Voxel Automata development for analyzing 3D densification and applying to point clouds: A synthetic review. Automation in Construction. 2018;**96**:236-249

[12] Sepasgozar S, Lim S, Shirowzhan S, Kim Y, Nadoushani ZM. Utilisation of a New Terrestrial Scanner for Reconstruction of As-built Models: A Comparative Study. ISARC. Proceedings of the International Symposium on Automation and Robotics in Construction. Vilnius Gediminas Technical University, Department of Construction Economics & Property. IAARC Publications; 2015. Vol.32. p. 1

[13] Sepasgozar SM, Forsythe P, Shirowzhan S. Evaluation of terrestrial and mobile scanner technologies for part-built information modeling. Journal of Construction Engineering and Management. 2018;**144**(12):04018110

[14] Sepasgozar SM, Forsythe P, Shirowzhan S, Norzahari F. Scanners

and photography: A combined framework. In: Singhaputtangkul N, editor. The 40th Australasian Universities Building Education Association (AUBEA) 2016 Conference. Cairns: Central Queensland University; 2016. pp. 819-828

[15] Sepasgozar SM, Wang C, Shirowzhan S. Challenges and opportunities for implementation of laser scanners in building construction. In: Sattineni A, editor. 33rd International Symposium on Automation and Robotics in Construction. Auburn, Alabama, USA: ISARC; 2016. pp. 742-751

[16] Shirowzhan S, Sepasgozar SM. Spatial analysis using temporal point clouds in advanced GIS: Methods for ground elevation extraction in slant areas and building classifications. ISPRS International Journal of Geo-Information. 2019;**8**:120

[17] Sepasgozar SM. Digital technology utilisation decisions for facilitating the implementation of Industry 4.0 technologies. Construction Innovation. 2020;**21**:476-489

[18] Sepasgozar SME, Davis S. Digital construction technology and job-site equipment demonstration: Modelling relationship strategies for technology adoption. Buildings. 2019;**9**:158

[19] Sepasgozar SM, Davis SR, Li H, Luo X. Modeling the implementation process for new construction technologies: Thematic analysis based on australian and us practices. Journal of Management in Engineering. 2018;**34**:05018005

[20] Sepasgozar SM, Davis S, Loosemore M, Bernold L. An investigation of modern building equipment technology adoption in the Australian construction industry. Engineering, Construction and Architectural Management. 2018;**25**:1075-1091

[21] Sepasgozar SM, Davis SR, Loosemore M. Dissemination practices of construction sites' technology vendors in technology exhibitions. Journal of Management in Engineering. 2018;**34**:04018038

[22] Sepasgozar SME, Loosemore M. The role of customers and vendors in modern construction equipment technology diffusion. Engineering, Construction and Architectural Management. 2017;**24**:1203-1221

[23] Niemeläa M, Shia A, Shirowzhana S, Sepasgozara SM, Liua C. 3D Printing Architectural Freeform Elements: Challenges and Opportunities in Manufacturing for Industry 4.0. ISARC; 2019

[24] Lim S, Le T, Webster J, Buswell R, Austin A, Gibb A, et al. Fabricating construction components using layered manufacturing technology. Global Innovation in Construction Conference. 2009:512-520

[25] Khoshnevis B, Dutton R. Innovative rapid prototyping process makes large sized, smooth surfaced complex shapes in a wide variety of materials. Materials Technology. 1998;**13**:53-56

[26] Doyle L, Brady AM, Byrne G. An overview of mixed methods research–revisited. Journal of Research in Nursing. 2016;**21**(8):623-635

[27] Sepasgozar SM, Davis S. Construction technology adoption cube: An investigation on process, factors, barriers, drivers and decision makers using NVivo and AHP analysis. Buildings. 2018;**8**:74

[28] Davtalab O, Kazemian A, Khoshnevis B. Perspectives on a BIM-integrated software platform for robotic construction through Contour Crafting. Automation in Construction. 2018;**89**: 13-23

[29] Moledina MMG, Pin GW, Enegbuma WI, Ali KN, Adenuga K. Building information modelling technological innovations in industrialised building systems cost estimation. In: 2017 International Conference on Research and Innovation in Information Systems (ICRIIS). IEEE; 2017. pp. 1-6

[30] Khosakitchalert C, Yabuki N, Fukuda T. The accuracy enhancement of architectural walls quantity takeoff for schematic BIM models. ISARC. In: Proceedings of the International Symposium on Automation and Robotics in Construction. IAARC Publications; 2018. pp. 1-8

[31] Jiangsu Sipai Instrument Co L. SP-LDE Concrete Flowmeter [Internet]. www.KuYiBu.com. Available from: http://www.kuyibu.com/c_jhsnyb/p9473598.html2014 [Accessed 16 June 2019]

Chapter 7

Transparent Vacuum Insulation Panels

Takao Katsura

Abstract

New, low-cost transparent vacuum insulation panels (TVIPs) using structured cores for the windows of existing buildings are proposed. The TVIP is produced by inserting the structured core, the low-emissivity film, and the adsorbent into the transparent gas barrier envelopes. In this chapter, the authors introduce the outlines, the design and thermal analysis method, the performance evaluation (test) method. Firstly, five spacers, namely peek, modified peek, mesh, silica aerogel, and frame, are selected as the structured core. The effective thermal conductivity of TVIPs with five different spacers is evaluated at different pressure levels by applying numerical calculation. The result indicated that TVIPs with frame and mesh spacers accomplish better insulation performance, with a center-of-panel apparent thermal conductivity of 7.0×10^{-3} W/m K at a pressure of 1 Pa. The apparent thermal conductivity is the same as the value obtained by the simultaneous evacuation thermal conductivity measurement applying the heat flux meter method. Furthermore, using a frame-type TVIP with a total thickness of 3 mm attached to an existing window as a curtain decreases the space heat loss by approximately 69.5%, whereas the light transparency decreases to 75%.

Keywords: transparent vacuum insulation panels, structured core, retrofitting insulation, design and numerical analysis, thermal conductivity measurement

1. Introduction

Environmental issues and the global energy needs have become the most crucial world concerns. This is because of the yearly increase in the annual energy consumption, in addition to the acceleration in the environmental pollution amount. Therefore, several energy-saving thermal insulation techniques have been applied for building applications in new established buildings. On the other hand, the old historical buildings are often protected due to their contributions to society. These thermal insulating technologies are rarely installed to the existing buildings. That is why the insulation capability of the existing buildings is not high. Consequently, it is a mandatory to enhance the insulation performance for the existing buildings. Vacuum insulation panels (VIPs) can be used for this aim. The VIPs have a very low thermal conductivity [1]. Therefore, they considered as a promising method for enhancing the thermal insulation of buildings, specifically for energy retrofitting (where higher insulation performance and smaller thicknesses are desirable) [2]. Further, it is indispensable to decline winter heat loss or heat gain in summer season for the windows of the existing buildings without much affecting the window transparency. The significant decrease in the window transparency increases the daytime energy consumption for the lighting purposes [3].

IntechOpen

The research on VIPs has been applied in some fields to improve insulation performance. For example, in some buildings, it can attain five times higher insulation values than conventional building insulating materials [4]. Recently, the common manufactured VIPs are applied for building walls insulation. These VIPs have an opaque appearance. These VIPs structures contain a metallized aluminum film. This aluminum envelope contains a high thermal insulation material under vacuum conditions and placed between the structure layers of a building wall.

Silica powder has been applied for the trial manufacture of VIPs. A silica aerogel with a thermal conductivity of 2×10^{-2} W/m K is usually used to make transparent VIPs [5]. It has been found that using such material in smart windows decreases the U-values by 63% as compared to conventional glass windows. In addition, a significant reduction in the light transmittance (by approximately 30%) has been measured [5]. Moreover, the industrial production cost of a VIP using silica aerogel remains very high. Hence, there is a need to search for another economically competitive material.

A new slim structured-core and transparent insulation method for windows of existing buildings is proposed. The TVIP is produced by inserting the structured core, the low emissivity film, and the adsorbent into the transparent gas barrier envelopes. In this chapter, the authors firstly describe the outlines of TVIPs. Next, the core structural design to maintain a vacuum layer and the thermal analysis method to predict the insulation performance are explained. In addition, measurement methods of apparent thermal conductivity and light transmittance, which are used to evaluate the insulation performance and transparency of TVIPs, are introduced. Furthermore, the results of thermal performance analysis and annual energy analysis are demonstrated.

2. Outlines of transparent vacuum insulation panels

2.1 Outlines, applications, and advantages

Figure 1 shows a concept diagram of the TVIP [6]. The manufacturing steps start with the preparation of the VIP core structure, transparent low-emissivity (Low-E) sheet, and transparent envelope. The heat sealing of the envelope is a four-sided seal type. In addition, the Low-E film coated on one face, whereas the other face has a larger emissivity. The core structure and the Low-E film are inserted in a transparent gas barrier envelope. In addition, the oxide calcium adsorbent pack or the getter material is also inserted to adsorb the outgassing from core material and the gas passing through the envelope. The transparent envelope facilitates the applicability of these TVIPs to existing windows where light transparency is

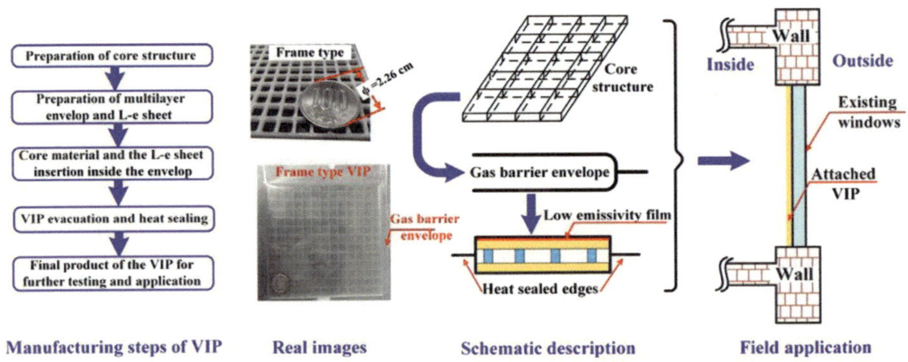

Figure 1.
Manufacturing steps, real images, schematic description, and the proposed real field application of the structured-core and transparent vacuum insulation panel (TVIP) for existing windows [6].

indispensable. The full structure is evacuated and sealed in a vacuum sealing machine after reaching the vacuum design pressure. **Figure 2** shows an example of lab-scale vacuum sealing machine [6]. In this machine, the pressure level can be controlled either automatically or manually. After reaching the desired pressure, sealing is automatically performed. Then, the sample can be taken from the vacuum chamber. The vacuum sealing machine usually consists of a few vacuum pumps (the vacuum sealing machine shown in **Figure 2** has buster and rotary vacuum pumps.). Also, the sealing can be achieved by a machine.

Usually, a sealing with approximately 10 mm width are used to prevent the penetration of atmospheric pressure. **Figure 1** depicts the real appearance, schematic illustration, and the method used for the field applications for the proposed VIPs. These VIPs might contribute to enhance the insulation capability of current existing windows, with slight effect on the daytime light transparency. These VIPs can be integrated as transparent curtains with low thermal conductivity and small thickness.

The advantage of TVIPs applying the structured core materials compared with the conventional transparent insulation technologies such as the vacuum glazing and the TVIP using silica aerogel is low cost to produce. **Figure 3** presents the comparison of costs [6]. The total TVIPs cost including envelope, inner spacers, and film of Low-E coating are shown in **Table 1** [6]. It is obvious that the cost of VIPs is around one-third of using double-glazing with vacuum. Further, the currently proposed VIPs can be attached to the existing windows with lower installation cost compared to silica aerogel VIPs. Finally, the cost of L-e film is around 50% of the total VIP price.

2.2 Core materials

The structure core materials are required the following items.

1. Strong enough to maintain a vacuum layer against atmospheric pressure from outside of envelope.

2. Low thermal conductivity to reduce the thermal bridge generated by the core materials.

3. No outgassing after vacuum sealing.

4. Easy to manufacture and inexpensive.

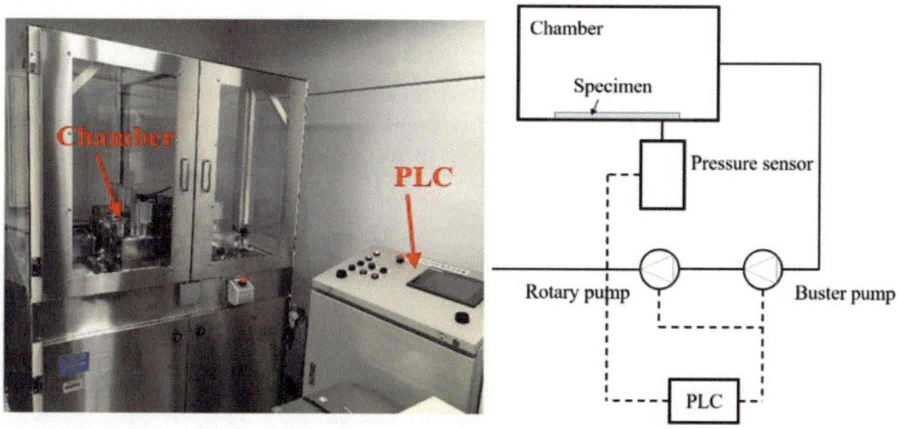

Figure 2.
Photograph and schematic of the vacuum sealing equipment [6].

Real photos					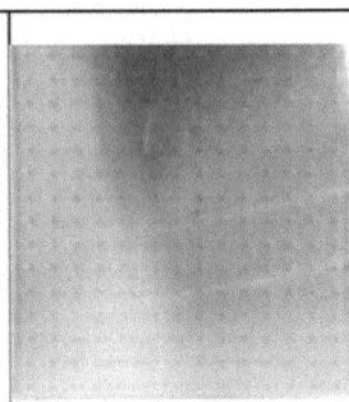
Design	**Double-layered vacuum glazing**	**Frame VIP**	**Mesh VIP**	**Silica aerogel VIP**	**Pillar supported VIPs**
Total Cost	**> 270 $ + (90 $ construction fee)**	**< 99 $**	**< 90 $**	**> 4500 $**	**< 135 $**

Figure 3.
Comparison of the calculated production cost of proposed VIPs with the double-layered vacuum glazing insulation [6].

Item	Mesh spacer	Frame spacer	Peek spacer	Silica aerogel
Core material	9 $	18 $	36 $ (plates + pillars)	4401 $
L-e film	45 $	45 $	45 $	45 $
Envelope	18 $	18 $	18 $	18 $
Manufacturing fee + adsorbent	18 $	18 $	36 $	36 $
Total cost	90 $	99 $	135 $	4500 $

Table 1.
Detailed costs of the manufactured VIPs ($\$/m^2$) [6].

The structured core materials used for TVIPs are depicted in **Figure 4** [6]. These TVIP designs include a peek spacer, mesh spacer, frame spacer, modified peek spacer, and silica aerogel spacer. All of these VIPs are designed, manufactured, and experimentally tested, except for the modified peek spacer. The modified peek spacer is proposed as extensions for the peek spacer. The TVIPs with peek spacers and aerogel spacers are superior to the TVIPs using a frame spacer and a mesh spacer from the viewpoint of transparency. On the other hand, the VIPs using a frame spacer and a mesh spacer have an advantage that it is easy to manufacture. In particular, if the frame spacer can be mass-produced at low cost, it is considered to be the most promising core material for the TVIP.

2.3 Envelopes

Metalized film barrier envelopes are commonly used in the conventional VIPs; however, they cannot be applied to the TVIPs. Transparent gas barrier films are applied as a covering envelope for the structured core. These envelopes are favorable for sealing and transparency. **Figure 5** depicts one example of detailed structure and dimensions of the transparent gas barrier envelope [6]. In order to reduce the oxygen and moisture transmission rates, the silica evaporated PET layer is applied.

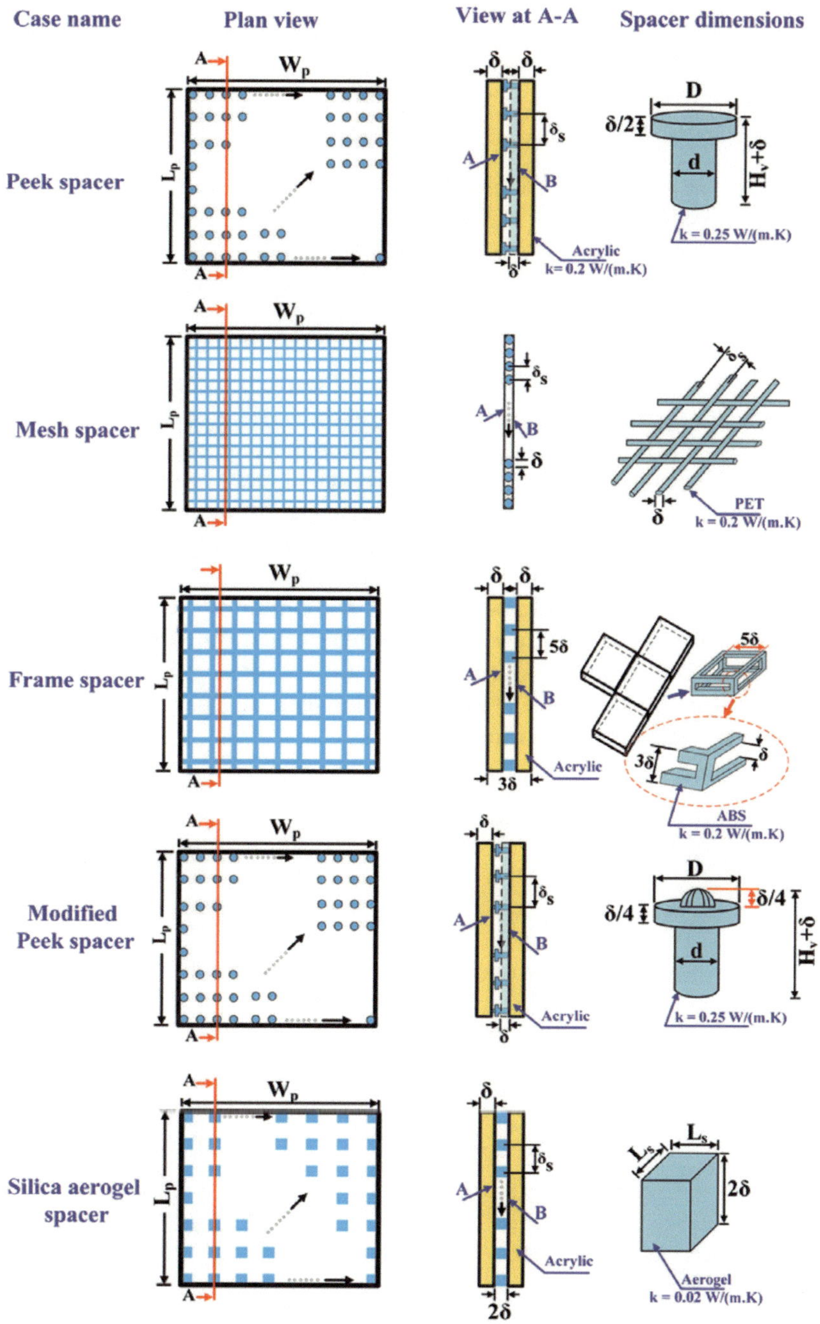

Figure 4.
Concept diagram of the proposed structured-core and transparent VIPs with different core structures of peek spacer, mesh spacer, frame spacer, modified peek spacer, and silica aerogel spacers [6].

3. Design and thermal analysis of transparent vacuum insulation panels

3.1 Core structural design

In order to determine the reasonable size of the VIP, the stress analysis is required. In this chapter, an example of stress analysis for frame spacer is

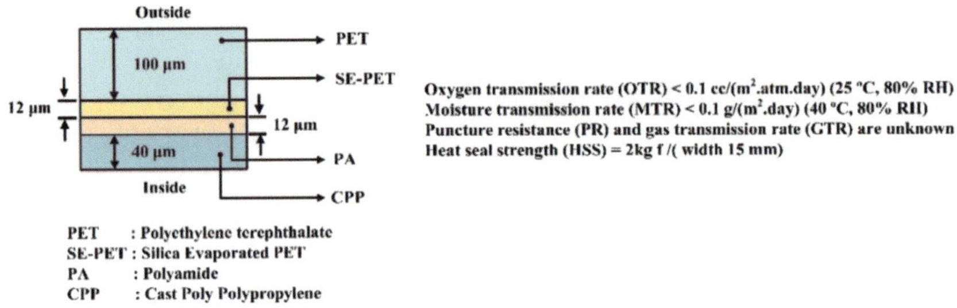

Figure 5.
One example of detailed structure of the transparent gas barrier envelope [6].

demonstrated. The structural frame depicted in **Figure 6** (on the left) [7] is used to maintain a vacuum gap when the internal pressure is decreased to 0.1 Pa. This clarifies that every frame should hold up an atmospheric pressure loading. The representation is shown in **Figure 6** (on the right) [8, 9]. The structural calculation analysis conducted to a model in **Figure 7** is separated into two possibilities.

The structural model used in this work is simplified as support beam. The deflection is used to estimate the structural dimensions. The calculation model shows the maximum stress could be happen at the center of the beam and while the minimum stress occurs at the frame edge. In this simulation, when the atmospheric pressure is used as the loading, a compression and elastic occur simultaneously but with opposite directions. Then, a relationship between the span and the deflection is defined using the following equation:

$$\overline{\omega}_{max} = \sigma \times \frac{P_0 \times a^4}{D} \quad (1)$$

$$D = \frac{Et^3}{12(1-\mu^2)} \quad (2)$$

If the polycarbonate frame is used, the Young's modulus E is 2.32 GPa and the Poisson's ratio μ is 0.39. The polycarbonate frame width, t, is used as 1 mm, while

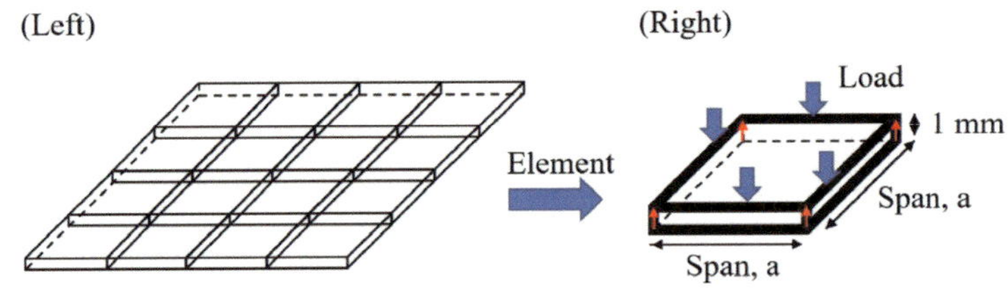

Figure 6.
One example of structure analysis for a frame-structured VIP [7].

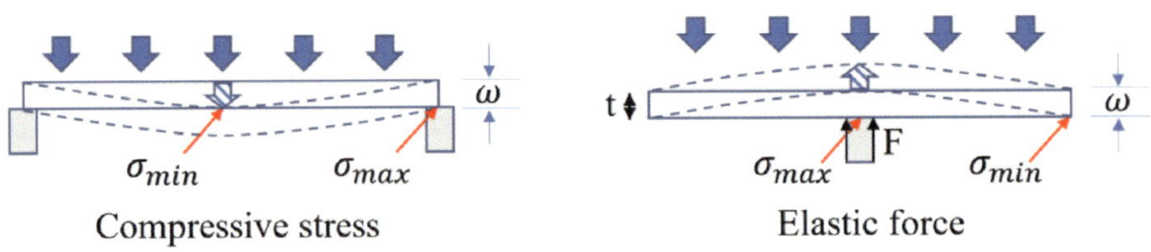

Figure 7.
Representation figures of the proposed structural model [7].

the span, *a,* is used as10 mm, the uniform loading coefficient to a rectangular flat plate σ_1 is 0.0138, and σ_2 is 0.0611. Further, the value of P_0 is used to be 1013 kPa. The frame extreme deflection with different spans vales is revealed in **Figure 8** [7]. However, the effect of these structural designs significantly influences the insulation capabilities. In order to create an appropriate design for VIP, we discussed additional models. Therefore, we designed the structure by changing the width of frame from 0.5 to 1.5 mm, and changing the span to 5, 8, 10, 12, and 15. The influence in deflection is revealed in **Figure 9** [7].

The frame mechanical structure is designed within the safety condition is showed in **Figure 8**. The frame span of 10 mm is used, while considering the stress effects and distortion, to used in the numerical model of thermal transmittance and to predict the thermal insulation capabilities of the proposed specimen in the experimental manufacturing. Furthermore, the vacuum gap was attained, and the frame surface was quite flat with a designed span of 10 mm. **Figure 9** compares the deflections at different conditions. The frame design also considered the

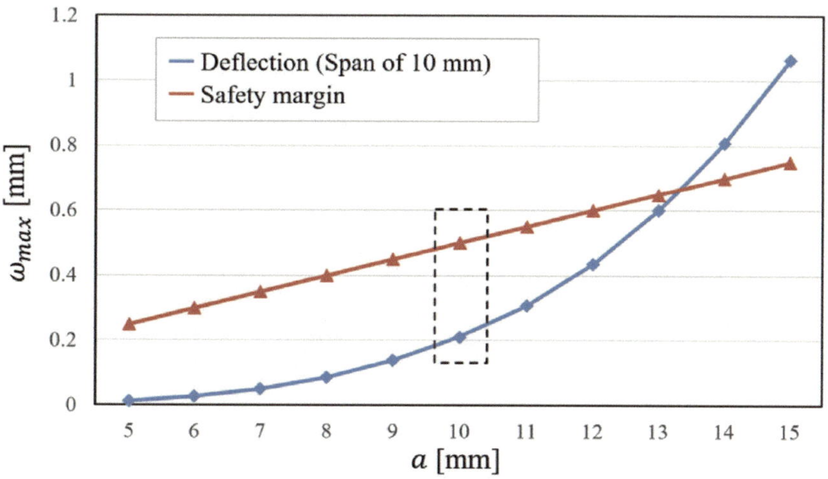

Figure 8.
The variation of maximum imposed deflection of the frame design at different span values [7].

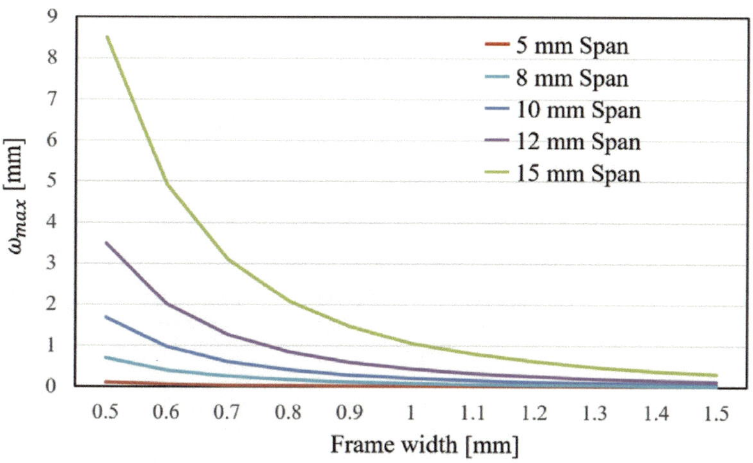

Figure 9.
Variation of the estimated deflections under various design conditions [7].

transparency along with the insulation performance. Hence, the frame with width of 1 mm, span of 10, and transparent gas barrier envelope is used.

To validate the aforementioned design, a 3D model was developed using ANSYS workbench 14.0. The distribution of deformation for the core frame that is modeled as concept in **Figure 10** and the result is shown in **Figure 11** [7]. The fixed boundary indicates that the element analysis can be applied to a full-scale VIP, and the meshing in **Figure 10** shows the calculation result is accurate due to the 257,791 nodes and 49,046 elements. The result validates the mechanical analysis in **Figure 8**, indicating a reasonable structural design with a sufficiently flat for application.

3.2 Thermal analysis applying numerical calculation

In VIPs, the inner pressure must be reduced to below 10 Pa. Increasing the gas pressure above this value rapidly increases the VIP thermal conductivity [10]. In addition, the total heat transfer in VIPs can be divided into four parts: radiation heat transfer through the vacuum space, heat conduction through the skeleton of the core of the core structure, gas conduction, and gas convection [10]. To enhance the insulation performance of the VIP, all of these parts must be minimized. However, at very low pressures, the convection can be naturally converted to pure gas conduction. Therefore, the convection can be ignored in the calculation [11–13]. Hence, the current problem contains a coupled conduction radiation heat transfer. The 3D

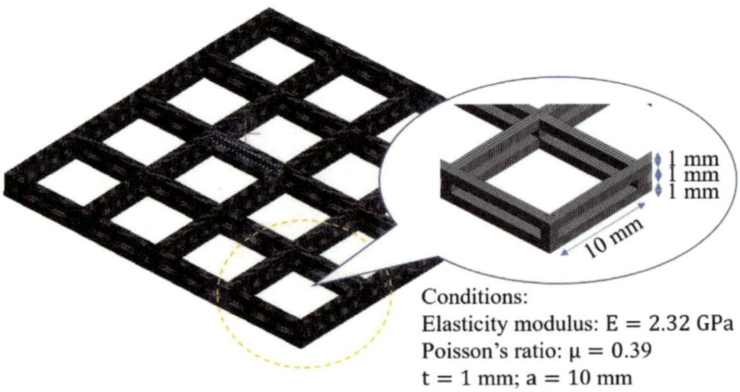

Figure 10.
Three-dimensional meshing stage in finite element model [7].

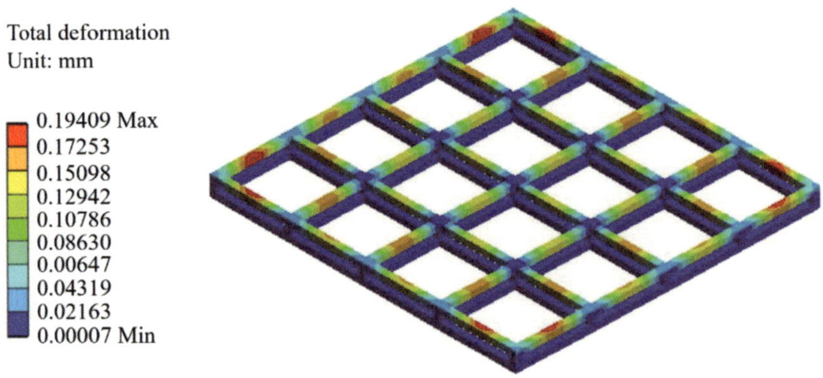

Figure 11.
Distribution of deflection on frame (three-dimensional) [7].

heat conduction equations for the solid regions including the pillars and in TVIP were coupled with the surface to surface radiation model. The current model adopts the following assumptions:

a. Steady state analysis is conducted.

b. The gas convection effect can be neglected within the simulated vacuum pressure [11, 12].

c. Because the low-emissive films have a very small thickness, the conduction heat transfer in this layer is neglected. But, the effect of its low emissivity is taken into consideration in the radiation heat exchange in the vacuum region.

d. The heat transfer by conduction in the envelope of the TVIP is also ignored.

e. The thermal contact resistances in between the layer structures are negated.

f. Isotropic materials were assumed for the complete structure of the TVIP.

The steady-state 3D energy equation with radiation source term can be written as follows as mentioned in ANSYS FLUENT [14]:

$$\nabla.(k\nabla T) + S = 0 \tag{3}$$

where T is the element temperature, k is the element thermal conductivity, and S is a source term. This term is added to consider the effect of the radiation exchange in the vacuum space [14]. This equation is solved for all solid domains but with zero radiation heat source. Further, the gas conduction can be neglected at lower pressure. But, in this study, the gas thermal conductivity variation with the pressure, gas mean temperature, and vacuum layer thickness is considered in the calculations [15]. In this work, the air thermal conductivity in the vacuum space at different pressure, temperature, and pore size is calculated as follows:

$$\lambda_v = \frac{\lambda_o}{1 + \frac{(1.07\times10^{-7})T}{l_v P}} \tag{4}$$

where T, l_v, and P are the gas temperature in K, vacuum layer thickness in m, and pressure of the gas layer in Pa respectively. Further, k_o is the thermal conductivity of the air gap at a reference temperature, which is about 0.026 W/m K.

The radiation exchange is predicted in the TVIP is estimated using the surface to surface (S2S) radiation model existing in ANSYS. This model defines the faces participating in the radiation heat exchange and the view factor is calculated using the surfaces separation, orientation, and the surface size. The surfaces participating in the radiation heat exchange are the surfaces in contact with the vacuum regions [14]. This view factor is identified using the design geometry of the TVIP. Further details, and the S2S model limitations exists in ANSYS theory guide [14].

The boundary conditions used in this study were depicted in **Figure 12** including the schematic representation, boundary conditions, and the meshing details for all the proposed TVIPs [6]. One side of the TVIP is kept at a temperature of 35.5°C, while the other was kept at a temperature of 10.5°C. The wall emissivity and temperature are well defined in the vacuum zones. But, for the spacers, only the faces' temperatures are defined, because the radiation exchange happens only in the vacuum regions. Furthermore, the peripheral edges of the TVIP were assumed to be isolated due to the

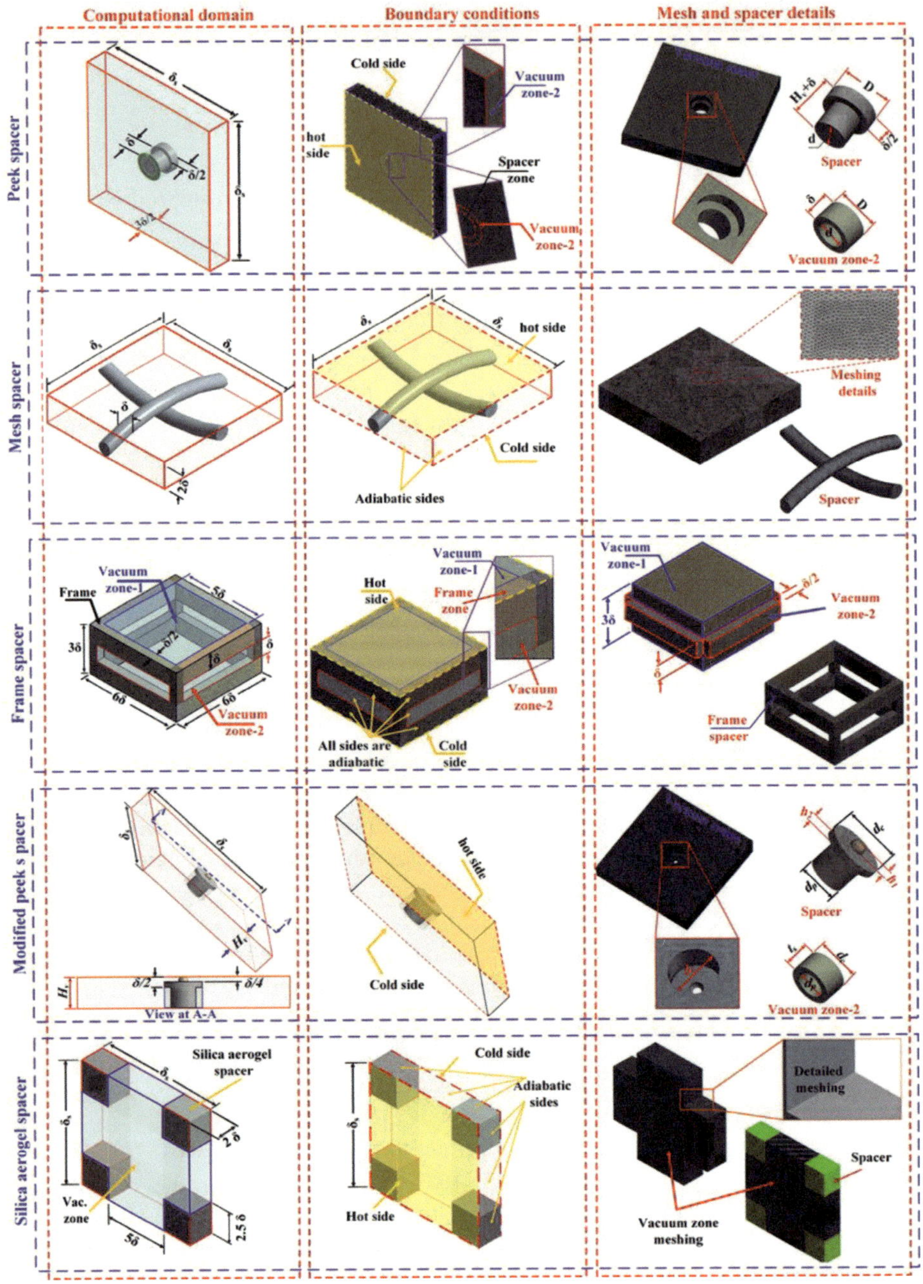

Figure 12.
Computational domains, boundary conditions, and mesh details for all investigated cases [6].

symmetrical computational domain. Finally, all interfaces in between the layers were thermally coupled. At this condition, the temperature and the heat transfer rate on these interfaces were assumed the same. The boundary conditions are mathematically expressed as follows for the for the frame-type spacer.

- For hot TVIP side:

  At the spacer face $T = T_{h.}$

  At Vacuum faces $= T_h$; $\varepsilon = \varepsilon_h$.

- For the cold wall:

  At the spacer zones $T = T_c$

  At the vacuum zones $= T_c$; $\varepsilon = \varepsilon_c$

Mesh study test is completed to confirm that the predicted results are not dependent on the mesh size. A total number of elements of 1,370,521, 2,450,133, 1,745,551, 864,000, and 1,600,000 were used for the simulation of the peek, modified peek, mesh-type, frame-type, and silica aerogel spacers, respectively.

First, the computational domains were created on the "DesignModular" software several zone approach. In this approach, the computational domain is separated into different zones. This allows us to control of element size, properties, and boundary conditions of every zone individually. Then, the generated domains are meshed as detailed in **Figure 12**. Fluent solver is used to solve the module governing equations. In this module, the flow equations including continuity, momentum, and the energy equation are the standard equations. At very low pressure, the convection effect can be neglected and therefore the fluid flow equations are deactivated [16–18]. But, the S2S model must solved for a fluid domain. So, the vacuum gap is simulated as a fluid zone, but only the energy equation is solved. The S2S model uses the estimated view factors. The energy equation for the solid zones and the vacuum space with the S2S radiation, are concurrently solved. Besides, the radiosity estimation is completed using the calculated view factors. The pressure effect is considered in the calculation by estimating the vacuum region thermal conductivity as a function of the vacuum pressure using Eq. (4). Layers for shell zone conduction are considered in the simulation to consider the effect of acrylic plates' existence.

The predicted center-of-panel thermal conductivity is compared with the measured thermal conductivity obtained by the Experiments for the peek type spacer,

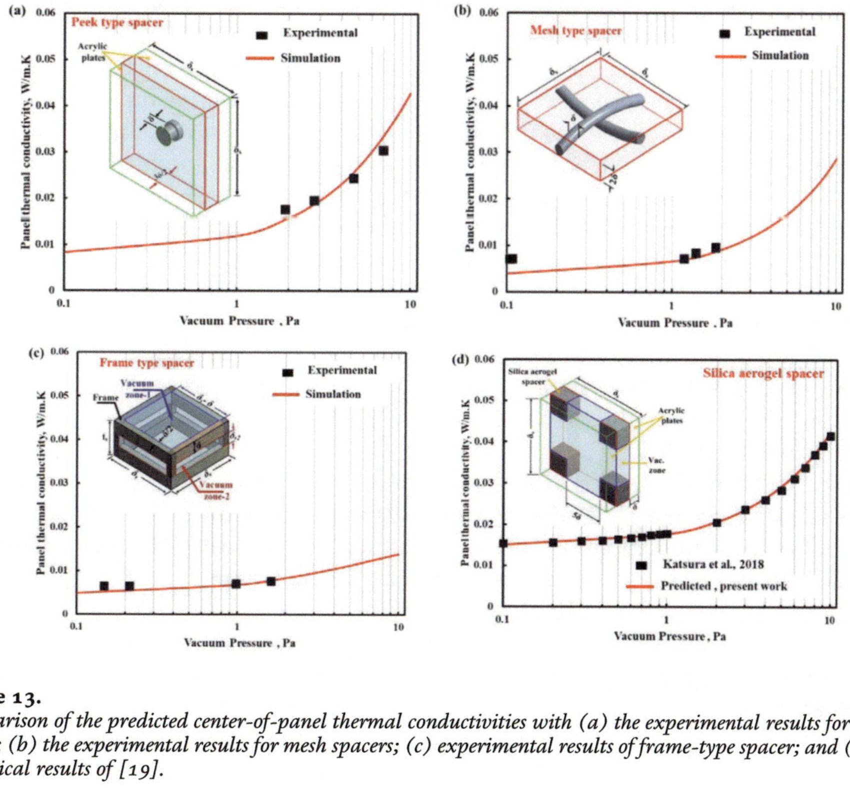

Figure 13.
Comparison of the predicted center-of-panel thermal conductivities with (a) the experimental results for peek spacer; (b) the experimental results for mesh spacers; (c) experimental results of frame-type spacer; and (d) the numerical results of [19].

mesh-type spacer, and frame-type spacer in **Figure 13a–c**, respectively. The experimental results in **Figure 13a–c** are obtained by the heat flux meter method (simultaneous evacuation thermal conductivity measurement) and the detail of experiment is described in Section 4.1.3. Further, the currently predicted results are compared with the numerical results of [19], as revealed in **Figure 13d**. The current model is inspected at pressure from 0.1 to 10 Pa. The predicted results in **Figure 13a–c** clarified that the model accurately predicts the center-of-panel thermal conductivity for the proposed TVIPs with different core structures with error less than 11%. This error is at lower pressure for the mesh-type spacer because of two reasons. The first reason is that at lower pressure, the contact between the rods become larger due to the flexibility of the rods. This reality is difficult to be considered in the simulation. The second reason may be attributed to the outgassing from the core structure.

Further, the current model is verified for the silica aerogel spacer with the results proposed in [19]. In this part, the silica aerogel spacer dimensions, boundary conditions, and vacuum zone dimensions are the same as those existing in [19], and are displayed inside **Figure 13d**. The main difference between the present model and the model developed by [19] is that the current model is a 3D model that uses S-S radiation, whereas the model developed in [19] was one-dimensional. From **Figure 13d**, a very good agreement is detected.

4. Performance evaluation method for transparent vacuum insulation panels

4.1 Thermal conductivity

Steady-state thermal conductivity measurement method for a flat material is used to evaluate the thermal insulation performance of TVIP. Two methods are commonly applied. The first one is the guarded hot plate (GHP) method and the second one is the heat flux meter (HFM) method. Of these two methods, the HFM method is recommended to use a standard sample with the same thermal conductivity as the material to be measured and to calibrate the heat flux meter. However, there is no standard sample with a thermal conductivity comparable to the VIPs. Therefore, the GHP method is said to be more suitable for high-precision measurements. Also, if the inside of GHP apparatus can be evacuated, it is possible to obtain the apparent thermal conductivity of TVIP for the set pressure. This means that the GHP method can provide the "ideal" apparent thermal conductivity of TVIP. Meanwhile, the HFM method has the advantage of easy measurement. In addition, the HFM apparatus is inexpensive compared to the GHP apparatus. From these reasons, in recent the HFM method is widely applied compared to the GHP method. Especially, the HFM method is selected when the speed and convenience are required such as quality control.

In the HFM method, TVIPs after vacuum sealing and TVIPs under evacuating can be used. When the TVIP after vacuum sealing is used and the HFM method is applied, the measured apparent thermal conductivity sometimes becomes higher compared to the expected value due to the outgassing from the core material. On the other hand, when the TVIP under evacuating is used and the HFM method is applied (simultaneous evacuation thermal conductivity measurement), it is possible to obtain the measured apparent thermal conductivity of TVIP for the set pressure as well as the GHP method.

In the following, the outlines and examples of each measurement method are introduced.

4.1.1 Guarded hot plate method

Figure 14 shows a schematic diagram of the GHP method. The two specimens, which have the same dimension and density, are positioned between the hot and cold plates and guarded in an airtight chamber. The cold plates are heated such that well-defined, selectable temperature difference ΔT°C is recognized between the two plates. The input heat power Q (W) through an area A (m^2) of the hot plate is measured at the thermal equilibrium conditions. Further, the input power, sample thicknesses, temperatures across the sample can be measured and hence the thermal conductivity can be determined at the steady-state conditions. Additionally, this method provides an evacuated space, which is accurate in evaluating the insulation capabilities in a stable vacuum situation. Until reaching the steady-state heat flow and vacuum level, the TVIP thermal conductivity, (λ), is estimated using the following equation:

$$\lambda = \frac{Q/2A}{\Delta T/d} \tag{5}$$

However, gaps are generated in the evacuated chamber in GHP method. Especially, if the shape of specimen is flat plate and the surface is hard, gaps are generated in the circled area in **Figure 14**. The gaps increase the thermal resistance; therefore, the thermal conductivity of TVIP evaluated by Eq. (5) becomes smaller due to the thermal resistance. Hence, the experimental result should be optimized using the following equation to eliminate the influence of the gaps:

$$R_{VIP} = R_m - R_{Va1} - R_{va2} - R_{va3} \tag{6}$$

where, R_{Va1}, R_{Va2}, and R_{Va3} are the thermal resistances of each gap. In addition, the apparent thermal conductivity of TVIP can be obtained by the following equation.

$$\lambda_{VIP} = L_{VIP}/R_{VIP} \tag{7}$$

An example of the thermal conductivity measurement applying GHP method, in which a frame structured core shown in **Figure 15** is used as the specimen, is demonstrated [7]. In this measurement, the inside the chamber of the GHP apparatus was reduced to 0.1 and 1 Pa. Then, the thermal conductivity was measured by using GHP method. The experimental results and their optimized results by Eq. (7) are shown in **Table 2** [7]. Furthermore, they are compared with the result of numerical calculation. **Figure 16** illustrates the comparison of apparent thermal

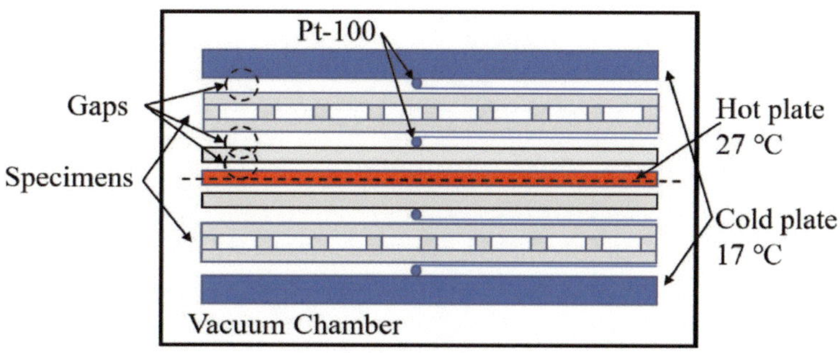

Figure 14.
Schematic of the guarded hot plate apparatus [7].

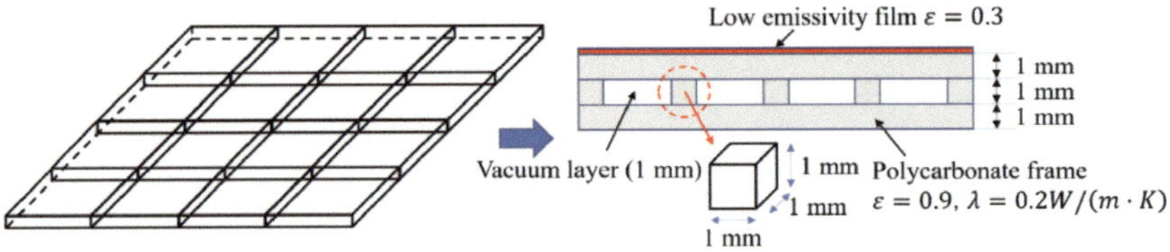

Figure 15.
Concept diagrams of the VIP specimen [7].

Pressure (Pa)	Experimental thermal conductivity λ_m (mW/(m·K))	Experimental Thermal resistance R_m ((m²·K)/W)	Thermal resistance of gaps R_{va} ((m²·K)/W)	Thermal resistance of VIP (Optimized) R_{VIP} ((m²·K)/W)	Thermal conductivity of VIP (Optimized) λ_{VIP} (mW/(m·K))
0.1	2.7	2.10	0.21	1.27	4.2
1	3.4	1.57	0.17	0.88	6.2

Table 2.
Guarded hot plate (GHP) experimental results [7].

conductivity of TVIP between experiment and calculation [7]. These results indicate that the optimized results are closer to the results of numerical calculation.

4.1.2 Heat flux meter method

Figure 17 shows a schematic diagram of the HFM method. The TVIP specimen is inserted between the hot plate and the cold plate. The temperature of the hot plate and the cold plate can be kept at constant. Then the heat flux is generated by the temperature difference between the hot plate and the cold plate and it is measured by the heat flux meter.

The apparent thermal conductivity of TVIP is measured based on a Fourier equation of one-dimensional heat conduction, as the following:

$$\lambda_{eff,exp} = \left(\frac{q_h + q_c}{2}\right) \times \frac{L_{VIP}}{\Delta T} \tag{8}$$

where the $\lambda_{eff,exp}$, q_h, q_c, L_{VIP}, and ΔT are the TVIP measured thermal conductivity in W/m K, heat flux measured at the hot and cold sides of the sample

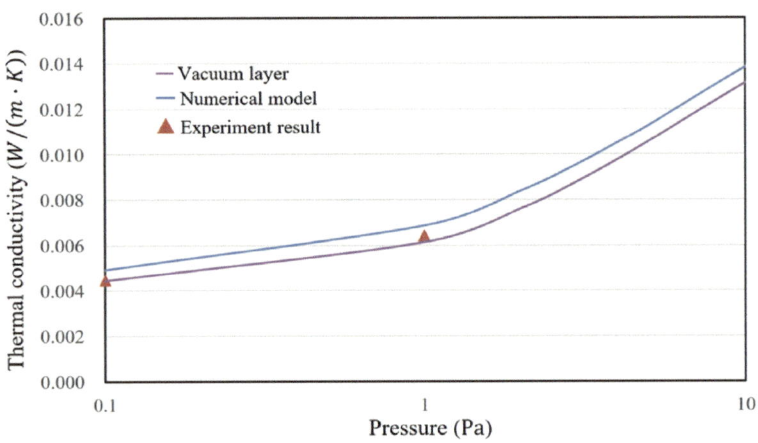

Figure 16.
Comparison of the simulation and experimental results [7].

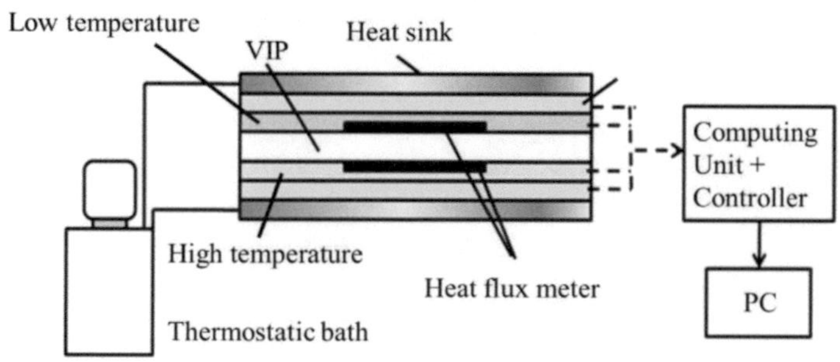

Figure 17.
Schematic diagram of heat flux method apparatus.

measured using two heat flux sensors in W/m^2, the TVIP thickness in m, and the difference in temperatures across the sample sides in °C, respectively.

An example of the thermal conductivity measurement applying HFM method is demonstrated [6]. Four types of core materials (peek spacer, mesh spacer, frame spacer, silica aerogel spacer) are used and the specifications are shown in **Figure 2** and the dimensions are indicated in **Table 3** [6]. The VIPs are produced by inserting the core materials, Low-E film (the emissivity is approximately 0.28), and oxide calcium adsorbent packs into gas barrier envelopes and vacuum sealing. The sealing was performed after the pressure inside chamber was lower than 0.1 Pa. Then, the apparent thermal conductivities of TVIPs were measured by using the heat flux method. **Figure 18** shows the measured apparent thermal conductivity according to elapsed time [6]. The apparent thermal conductivity is increasing according to elapsed time. This issue should occur due to the increased pressure, and the reason should be due to the outgassing from the core material.

In addition, the core material outgassing is minimized in different methods. First, **Figure 19a** [6], called "normal case," the TVIP is assembled by the vacuum sealing machine in **Figure 2**, and a calcium oxide powder placed in an adsorbent pack is used to reduce the core outgassing. Second, in **Figure 19b** [6], called "vacuum drying with N_2," all the TVIP components are placed inside an environmentally controlled chamber. In this chamber, the temperature was maintained constant at 70°C for a period of 24 h. After that, the TVIP structure is evacuated using another vacuum machine with turbomolecular vacuum pump to keep the pressure at 0.1 Pa. This could enhance the release of the outgassing from the core material of the TVIP. Then, nitrogen is supplied to the chamber then the nitrogen is exhausted. Finally, the TVIP is evacuated and evacuated and sealed using the machine in **Figure 2**.

Dimension	Value (mm)	Dimension	Value (mm)
δ	1	D	1.8
W_p	150	L_p	150
d	1.2	H_v	1.5
δ_s	10	L_s	5

Table 3.
Detailed dimensions of the current structured-core and TVIPs [6].

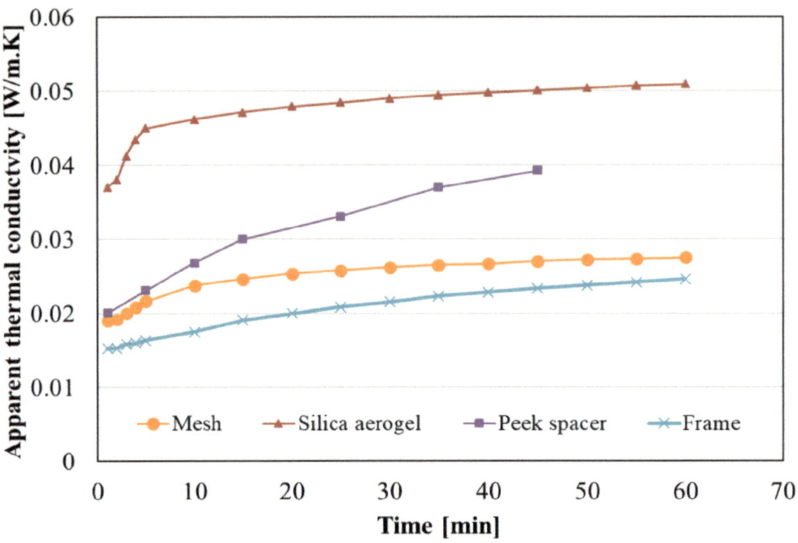

Figure 18.
Measured apparent thermal conductivity according to elapsed time [6].

Third, **Figure 19c** [6], called "heating," a new method is introduced to decrease the production processes used for the fabrication of the TVIP. In this method, heating up the VIP is conducted during the evacuation using two flat silicon rubber heaters. This helps the outgassing during the evacuation. The heaters temperatures were kept at 70°C using controller. In both second and third methods, calcium oxide adsorbent packs were kept inside the TVIP. Finally, in the fourth method displayed in **Figure 19d** [6], called "gas adsorbent getter," a different absorbent material is used. This material contains calcium oxide with alloy getter. The calcium oxide is used to absorb the water vapor and the alloy getter is used to absorb other gases such as nitrogen and carbon dioxide. Therefore, the pressure increase due to the outgassing could be decreased.

In **Figure 20a** [6], the predicted thermal conductivity variation with time is displayed for the four trial production methods. The comparison is executed for the TVIP with frame type. It is found that the use of getter material during the fabrication process achieves the lowest thermal conductivity compared to other methods. Further, manufacturing the VIP without using calcium oxide is the worst option. Because using the getter material attained the best results, further fabrication method is applied here. The use of getter is compared with the use of a calcium

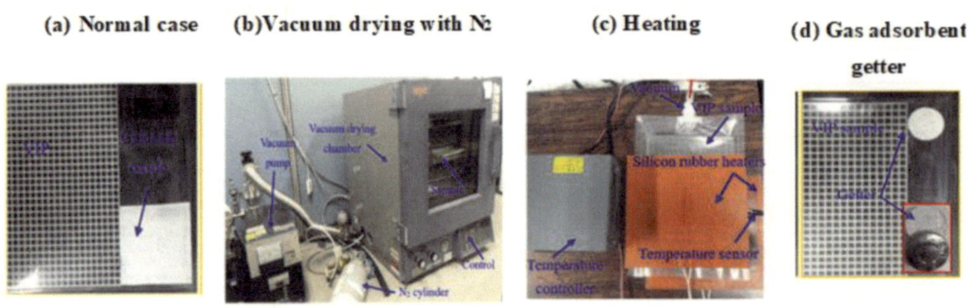

Figure 19.
Trial manufacturing methods for the frame-type VIP (a) normal method with calcium oxide adsorbent, (b) using vacuum drying chamber with N_2, (c) using flat silicon rubber heaters during the evacuation, and (d) using gas adsorbent getter [6].

oxide pack. The results were recorded for 75 h and are showed in **Figure 20b** [6]. The results showed that using getter material attained a lower panel effective thermal conductivity compared to using calcium oxide desiccant packs. The

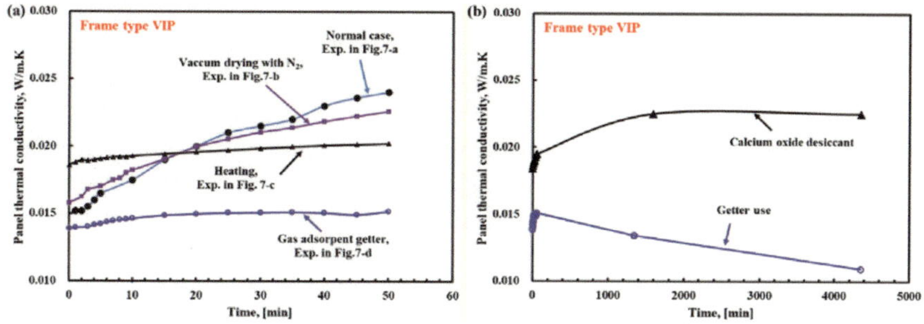

Figure 20.
Variation of the center-of-panel thermal conductivity with the elapsed time for (a) different trial manufacturing methods and (b) two different outgassing adsorbent materials [6].

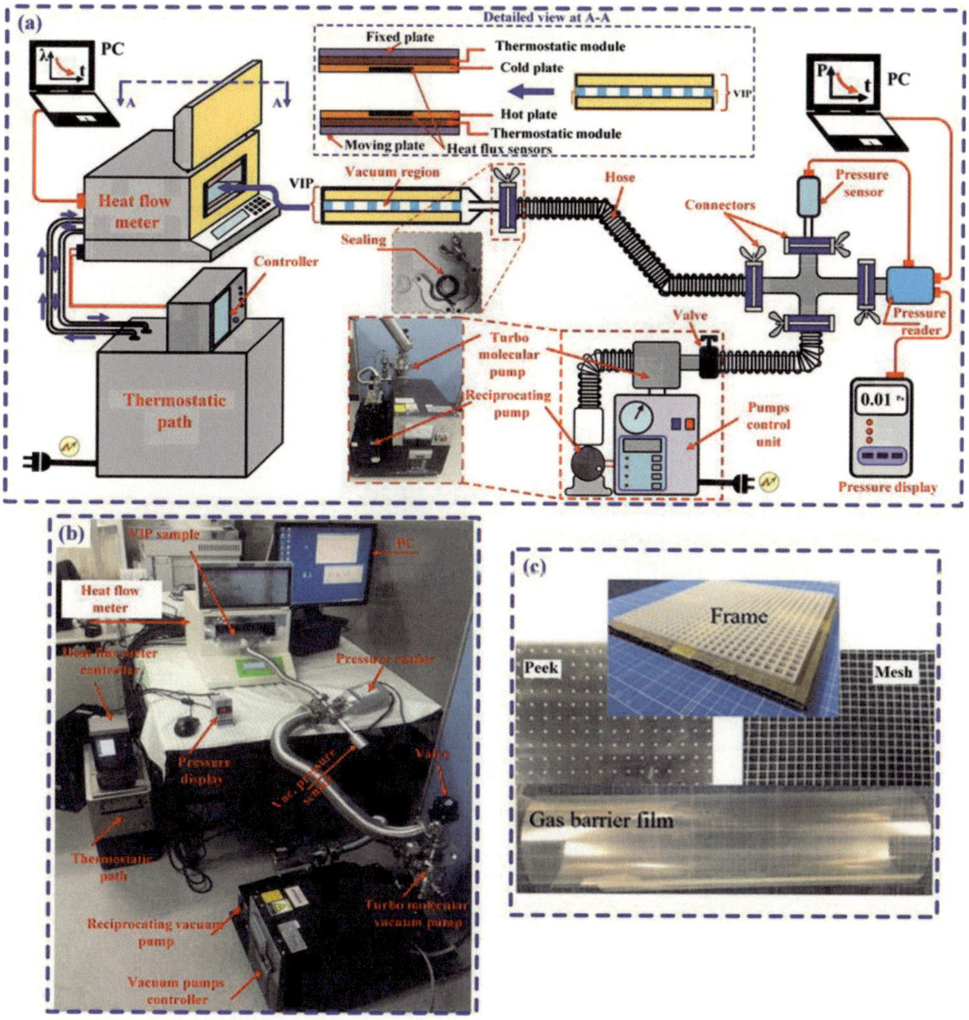

Figure 21.
Experimental apparatus used in the estimation of the VIP thermal conductivity: (a) Schematic diagram, (b) real images of the used test apparatus, and (c) real images of the used peek, frame, and mesh VIPs and the transparent gas barrier film [6].

minimum achieved thermal conductivity was about 0.011 W/m K, and this value corresponds to 5 Pa in the case of using getter. This clarifies that the VIP inner pressure is increased after the sealing, because of core material outgassing.

4.1.3 Heat flux meter method (simultaneous evacuation thermal conductivity measurement)

Figure 21 shows the simultaneous evacuation thermal conductivity measurement applying HFM method [6]. This technique achieves evacuation with simultaneous measuring of the thermal conductivity. In this experimental setup, the vacuum pump is connected with the experiments of the heat flow meter (HFM). Three sides of the VIP were sealed thermally using the large width sealing in the vacuum sealing machine. The width of the sealing was 9 mm. Then, fourth edge of the envelope is kept connected to the hose/pipe of a small-scale vacuum pump. The experimental setups are showed in **Figure 21**. The vacuum pump in this setup consists of two vacuum pumps. One is a series connected reciprocating and turbomolecular vacuum pump. Pirani pressure gauge is used to measure the VIP inner pressure. The VIP and the sensor used for pressure measurement can be easily isolated from the system by the mean of valve. The experiment starts with operating the reciprocating pump to

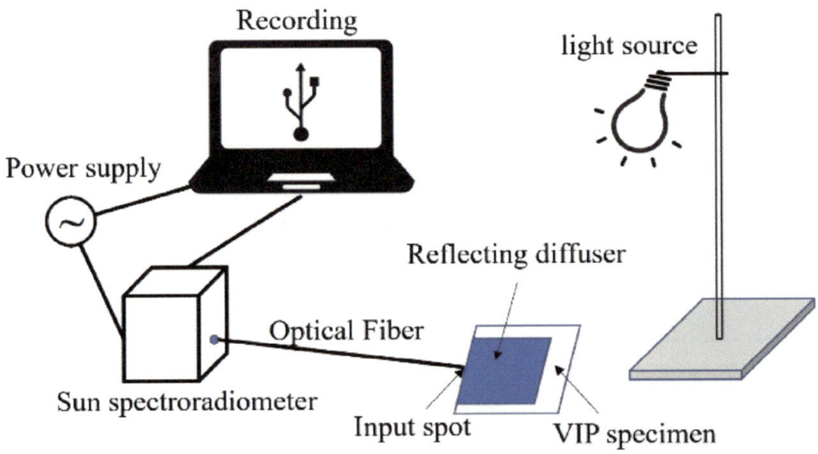

Figure 22.
Schematic representation for the illuminance measurement [6].

Conditions	Without VIP	Silica aerogel spacer	Peek type spacer	Mesh-type spacer	Frame-type spacer
Intensity ($\mu W/cm^2/nm$)	20000	14000	17500	13000	15000
Transparency Ratio	1	0.70	0.88	0.65	0.75

Table 4.
Measured transparency for the experimentally examined VIPs [6].

decrease the VIP inner pressure to around 20 Pa. Then, the turbomolecular pump is operated to decrease the inner pressure until reaching the desired value. The turbomolecular vacuum pump with the oil free type is favorable for these applications, as suggested by [18]. The pressure sensing device was located closer to the inlet part of the VIP; this allows us to precisely measure the pressure inside the VIP. After reaching the desired vacuum level, the VIP sample is inserted into the HFM apparatus. The measurement of thermal conductivity is the same as the normal HFM method. Examples of experimental results are indicated in **Figure 13a–c**.

4.2 Transparency

The light transparency experiment is described in **Figure 22** [6]. In this experiment, the TVIP sample is positioned on a reflection diffuser. Then the TVIP sample is irradiated using incandescent light lamp. An 800-nm wavelength was measured and the illuminance intensity in $\mu W/(cm^2\ nm)$ is measured with and without the TVIP shading. The transparency percentage is defined as the ratio between the measured value of the transmitted light intensity while using the TVIP the same value while not including the TVIP.

Table 4 shows the transparency percentage of all the fabricated TVIPs [6]. These transparency percentage is the TVIPs without the glass window of 3 mm. It is obvious that the transparency values for the manufactured VIPs vary between 0.65 and 0.9, for mesh-type and the cylindrical-pillar VIPs, respectively. Further, because the modified peek spacer was studied numerically, the predictable transparency could be similar to that of the peek type spacer.

5. Additional performance analysis of transparent vacuum insulation panels

5.1 Thermal performance analysis under American Society for Testing and Materials (ASTM) boundary conditions

A 3D model is developed and solved for the investigated TVIPs while they attached to a 3 mm glass layer thick. The temperature contours on both the cold and hot sides, heat flux, and U-value of the VIP samples were predicted at the ASTM boundary conditions. In ASTM boundary conditions, the indoor air temperature and the convection heat transfer coefficients were 21.1°C and 8.3 W/m^2 K while the same for the outdoor conditions were −17.8°C and 30 W/m^2 K, respectively.

The 3D temperature contours on the vacuum region at a vacuum pressure of 0.1 Pa are depicted in **Figure 23. Figure 23** shows a considerable temperature difference exists among the two sides of the vacuum gap. The maximum temperature difference was for the mesh-type spacer. Further, using the mesh-type VIP with 3-mm glass decreases the center of pane U-value from 6.3 to 1.14 W/m^2 K. This considerable decrease in the U-value declines the heat loss from the indoor environment of the building. Furthermore, using silica aerogel-based VIP, higher U-value was obtained. This due to the high heat is transfer rate through the silica aerogel spacer.

5.2 Annual energy analysis

The hourly variation and total annual thermal energy transferred through the proposed VIP to the indoor environment is calculated, analyzed, and compared with the customary window without VIP. The VIP is attached to the 1 m^2

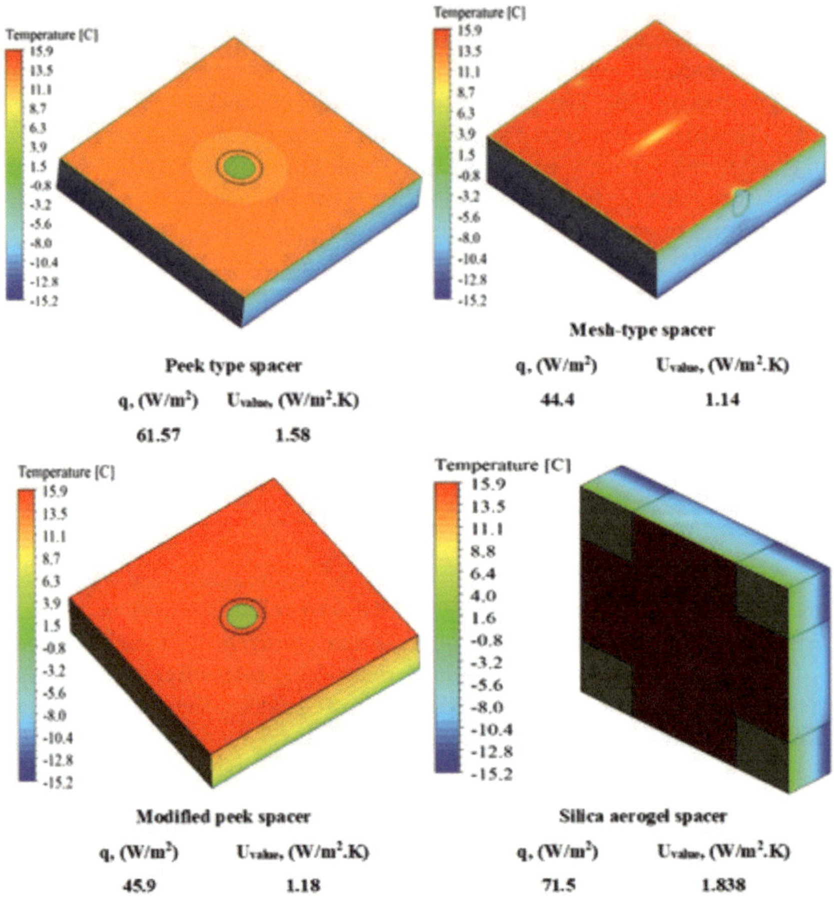

Figure 23.
Predicted 3D temperature contours of the investigated VIPs using American Society for Testing and Materials (ASTM) boundary conditions [6].

single-layered glass window with 3 mm thickness. In this calculation, the indoor air temperature is at 20°C, with an indoor convective heat transfer coefficient (HTC) of 7.7 W/m^2 K as recommended by Refs. [18, 20]. In addition, two typical climate conditions at Ishigaki island (24.41°N, 124.17°E) as a hot region, and Sapporo, Hokkaido (43.06°N, 141.35°E) as a cold region are examined in Japan. **Figure 20a** and **b** show the hourly variation of ambient temperature and wind speed, respectively. These meteorological conditions at Ishigaki city are measured by the authors, and that for Sapporo was downloaded from Japan Meteorological Agency website. In-house MATLAB code is developed to estimate the hourly and the annual thermal energy transfer. The minimum ambient temperature in Sapporo city is −10°C while the maximum air temperature in Ishigaki island is 32°C.

The model adopted in this section is a one-dimensional quasi-steady model. This model is an efficient for a long-term energy evaluation, as it predicts results with reasonably accepted accuracy in a short period of time.

In this model, the following assumptions are applied:

1. The heat transferred in one-dimensional and under a quasi-steady state condition.

2. The effect of the contact thermal resistance between the VIP and the glass window is neglected.

3. The impact of sol-air is not considered as we assume that no solar radiation strikes the exterior surface of the window.

The steady-state conduction heat flux transfer through the VIP-glass window system to the indoor space is calculated by Eq. (9):

$$q = \frac{(T_a(t) - T_i)}{\left(\frac{1}{h_{c,i}} + \frac{\delta_g}{k_g} + \frac{\delta_v}{k_v} + \frac{1}{h_{c,o}(t)}\right)} \tag{9}$$

where, q, δ_g, k_g, δ_v, k_v, and $h_{c,i}$ are the total thermal heat flux in W/m^2, glass thickness in m, glass thermal conductivity in W/m K, VIP thickness in m, VIP thermal conductivity in W/m K, and the inside convective heat transfer coefficient in W/m^2 K, respectively. Please note that the negative sign of q indicates that the heat loss from the building, which occurs when the outside temperature is less than the indoor designated temperature. The outdoor convective HTC variation with time is calculated as a function of wind speed and ambient temperature according to ambient temperature-dependent equation derived by Nusselt-Jürges [21]. This equation is written for smooth surfaces, such as the exterior surfaces of windows, as shown in Eqs. (10) and (11) [21]:

$$h_{c,o}(t) = 5.678\left\{0.99 + 0.21 \times \left[\left(\frac{294.26}{273.16 + T_a(t)}\right) \times \frac{U_\infty(t)}{0.3048}\right]\right\} \; at \; U_\infty(t) < 4.88\left(\frac{m}{s}\right) \tag{10}$$

$$h_{c,o}(t) = 5.678\left\{0.5 \times \left[\left(\frac{294.26}{273.16 + T_a(t)}\right) \times \frac{U_\infty(t)}{0.3048}\right]^{0.78}\right\} \; at \; 4.88 \le U_\infty(t) < 30.48\left(\frac{m}{s}\right) \tag{11}$$

Here, $h_{c,o}(t)$, T_a (t), and $U_\infty(t)$ are the hourly convective HTC at the exterior surfaces of the window in W/m^2 K, the hourly ambient temperature variation with the time in °C, and the hourly wind speed in m/s, respectively. The constants in Eqs. (10) and (11) are empirical constants derived from the experiments developed by Nusselt-Jürges. **Table 5** lists the thicknesses of the VIPs with the relevant thermal conductivities at a pressure of 1 Pa for different spacers [6].

The hourly variation of the heat flux for the single-layered glass window without/with a frame-type VIP attached to the internal surface of a glass window is shown in **Figure 24a** and **b** for Sapporo and Ishigaki, respectively [6]. These two figures show that attaching the VIP to the single-layer glass windows of existing buildings can reduce the heat loss/gain from/to the building in cold and hot weather conditions, respectively. Hence, it is highly recommended to use the VIP in cold regions such as Sapporo, where the temperature difference between the outside ambient temperature and the inside designed conditions can reach to maximum value of 30°C.

	Peek spacer	Mesh spacer	Cylindrical pillar	Frame spacer	Modified peek spacer	Silica aerogel
Thickness, (mm)	4.5	2	4.5	3	4.5	4
k_v @1 Pa, (mW/m.K)	14.2	6.5	16.7	6.8	12.4	17.7

Table 5.
Detailed thickness and center-of-panel thermal conductivities of the investigated VIPs at 1 Pa [6].

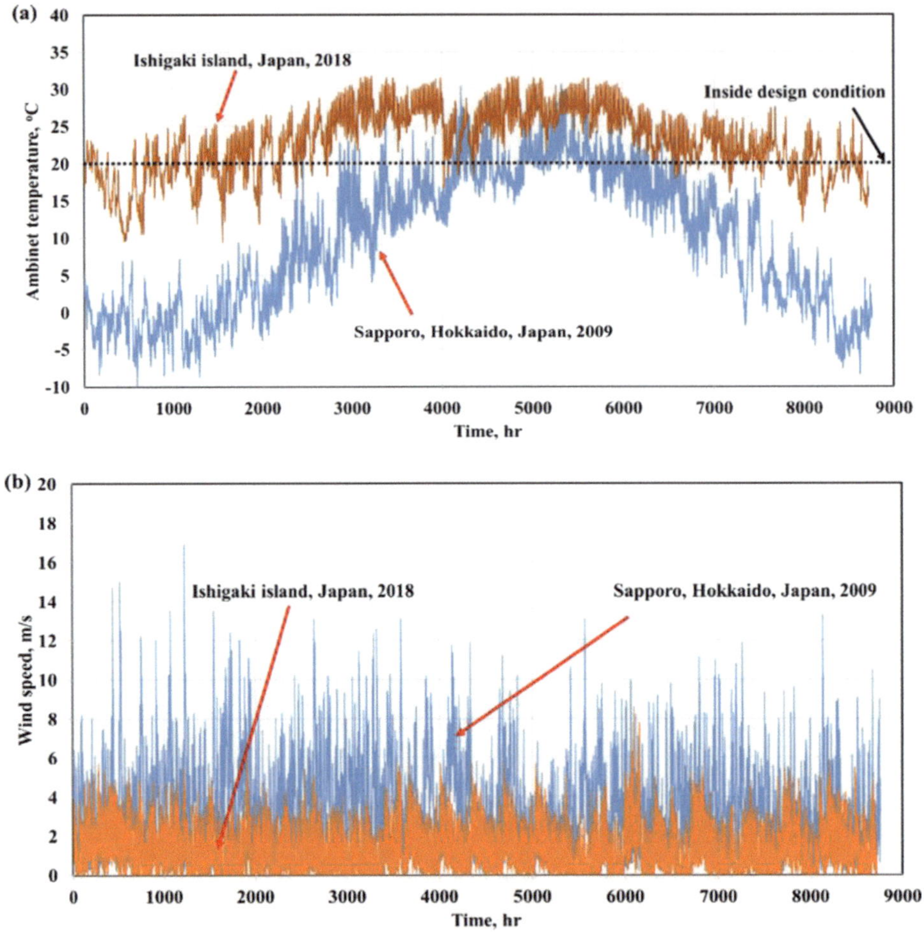

Figure 24.
Hourly variation of (a) ambient temperature and (b) wind speed for Ishigaki and Sapporo cities in Japan [6].

The hourly variation of the thermal heat flux in W/m^2 for a single layer window with/without VIP under climate conditions of Ishigaki Island and Sapporo city are displayed in the **Figure 25**, While, the total thermal energy transfer per unit floor area per annum in kW h/m^2/annum is calculated and compared for all VIPs with different spacer structure is illustrated in **Figure 26a** and **b** for Sapporo and Ishigaki, respectively [6]. First, the total annual heat flux in W/m^2/annum is the summation of loss and gain heat transfer. For the Sapporo region, it is obvious that the total annual heat flux loss is dominant heat transfer occurred in this region, owing to the cold weather. In addition, attaching a frame VIP to the inside surface of a customary 3-mm glass window reduces the heat loss from the buildings by 69.5% compared with the typical window without VIP. In contrast, for Ishigaki region, the heat gain from the window has a significate share of the total heat transfer rate, and a 65.1% decrement in the annual heat gain is accomplished by using the VIP combined with 3-mm glass windows.

6. Summary

New, low-cost transparent vacuum insulation panels (TVIPs) using structured cores are proposed. These TVIPs are different from the well-known conventional

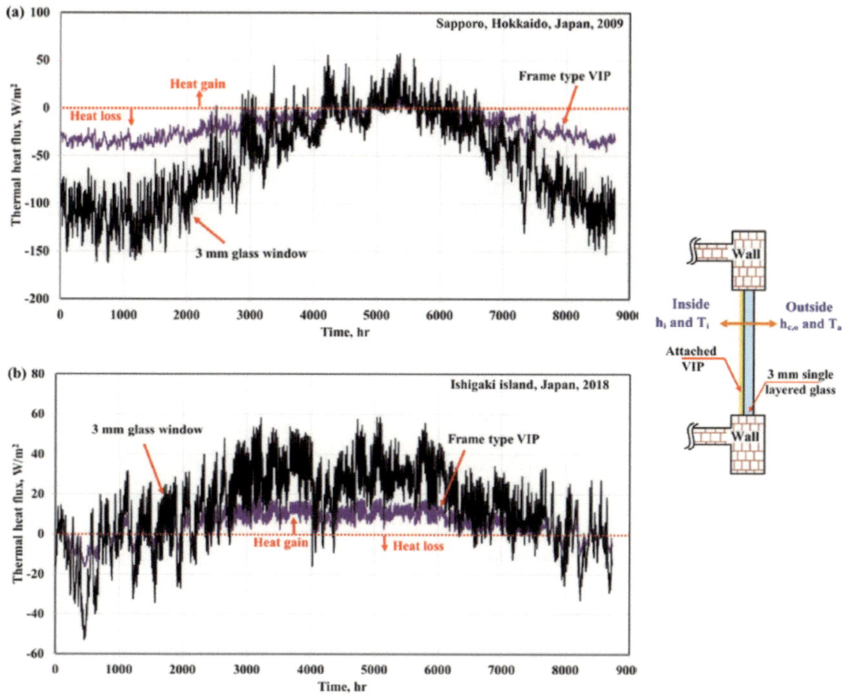

Figure 25.
Hourly variation of the thermal heat flux transfer in W/m² for (a) Sapporo and (b) Ishigaki in the case of using single-layered glass window and the frame-type VIP attached to the glass window as an example. The VIP with the glass window is displayed at the right of the figure [6].

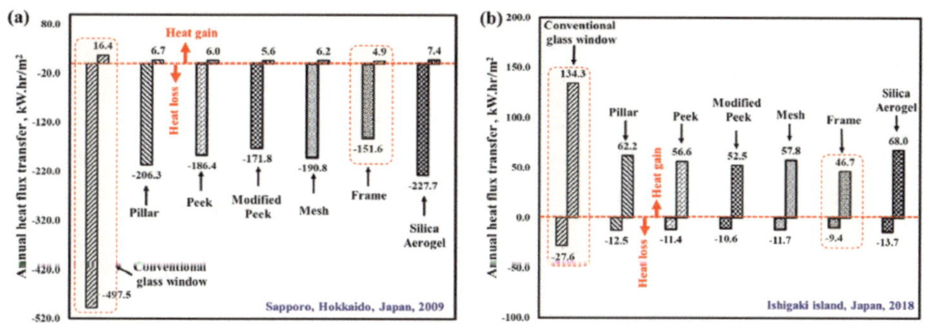

Figure 26.
Variation of the total annual heat flux transfer in a unit of kW h/m² for (a) Sapporo and (b) Ishigaki region [6].

VIP, which has a solid core material structure and opaque metalized film barrier envelope. The proposed novel TVIPs are showed a high thermal insulation capability, which could attached to the windows of existing buildings. The five spacers, which are namely peek, modified peek, mesh, silica aerogel, and frame, are selected as the structured core. Numerical simulations are developed to evaluate the effective thermal conductivity of TVIPs with five different spacers at different pressure levels, and the results are compared to the experimental pairs. The results indicate that the frame-type spacer with a thickness of 3 and 2 mm attain the lowest center-of-panel thermal conductivities of 6.5×10^{-3}, it followed by the mesh-type spacer, which achieved a value of 6.8×10^{-3} W/m K at a vacuum pressure of 1 Pa, respectively. The predicted U-values for the mesh spacers was 1.14 W/m^2 K at ASTM boundary conditions. Moreover, attaching a frame-type TVIP with a thickness of 3 mm to an existing window decreases the space heat loss by approximately 69.5%.

The TVIPs have the significant potential to promote thermal insulation capability of the windows of existing buildings with low cost. On the other hand, the challenge of TVIPs is to keep the low pressure inside the envelope after vacuum sealing. As shown in the result of thermal conductivity measurement applying HFM method and using a TVIP after vacuum sealing, the apparent thermal conductivity is increasing according to elapsed time. The reason should be due to the outgassing from the core material; therefore, it is important to reduce the outgassing to realize the TVIPs.

Acknowledgements

This work was financially supported by the Grants-in-Aid for Scientific Research of the JSPS (Research representative: Takao Katsura). The author thanks Prof. Katsunori Nagano, Dr. Ali Radwan, Dr. Zhang Yang, Mr. Makoto Nakamura, Mr. Masahiro Aihara, and Ms. Ririko Noda in Graduate School of Hokkaido University, Mr. Yuji Mori in Faculty of Engineering, Hokkaido University, Hokkaido Electric Power Co. Inc., Mitsubishi Chemical Corporation, Teijin Frontier Co., Ltd., ULVAC Inc., and Miyoshi Vacuum Packing for their assistance in the VIP test production.

Conflict of interest

The authors declare no conflict of interest.

Nomenclature

A	vacuum insulation panel front area (m^2)
a	length of span (m)
E	emissive power density (W/m^2),
E_y	Young's modulus (Pa)
$h_{c,i}$	convection heat transfer coefficient to building interior [$W/(m^2 K)$]
$h_{c,o}$	convective heat transfer coefficient to building exterior [$W/(m^2 K)$]
k	thermal conductivity [W/(m K)]
L_v	transparent VIP thickness (m)
P	vacuum pressure (N/m^2)
R	thermal resistance [(m^2 K)/W]
q	heat flux (W/m^2)
T	temperature (°C)
U_∞	wind speed (m/s)

Greek symbols

Δ	difference
δ	thickness (m)
ε	emissivity (unitless)
λ_o	air thermal conductivity [W/(m K)]
λ_v	vacuum space thermal conductivity [W/(m K)]

μ	Poisson's ratio
σ	coefficient of uniform loading to a rectangular flat plate
ω	deflection (m)

Subscripts

a	ambient
avg	average
c	VIP cold side
eff	effective
g	glass
h	VIP hot side
h-c	between the hot and the cold side of the VIP
i	interior
in	entering the surface
m	measured
out	exiting from the surface
VIP	VIP
v	vacuum space
v_1, v_2, v_3	vacuum layer

Author details

Takao Katsura
Hokkaido University, Sapporo, Japan

*Address all correspondence to: katsura@eng.hokudai.ac.jp

References

[1] Fricke J, Heinemann U, Ebert HP. Vacuum insulation panels—From research to market. Vacuum. 2008;**82**: 680-690. DOI: 10.1016/j.vacuum.2007.10.014

[2] Boafo FE, Kim J-H, Kim J-T. Numerical study of slim curtain wall spandrel with integrated vacuum insulation panel: Concept, performance evaluation and challenges. Energy and Buildings. 2019;**183**:139-150. DOI: 10.1016/j.enbuild.2018.10.036

[3] Yang Z, Katsura T, Aihara M, Nakamura M, Nagano K. Development of numerical heat transfer and the structural model to design slim and translucent vacuum layer type insulation panels to retrofitting insulation in existing buildings. Energies. 2017;**10**. DOI: 10.3390/en10122108

[4] Fantucci S, Garbaccio S, Lorenzati A, Perino M. Thermo-economic analysis of building energy retrofits using VIP-vacuum insulation panels. Energy and Buildings. 2019;**196**:269-279. DOI: 10.1016/j.enbuild.2019.05.019

[5] Buratti C, Moretti E, Zinzi M. High energy-efficient windows with silica aerogel for building refurbishment: Experimental characterization and preliminary simulations in different climate conditions. Buildings. 2017. DOI: 10.3390/buildings7010008

[6] Katsura T, Radwan A, Yang Z, Nakamura M, Nagano K. Energy conservation using new structured-core and transparent vacuum insulation panels: Numerical simulation with experimental validation. Solar Energy. 2019;**193**:885-905. DOI: 10.1016/j.solener.2019.09.083

[7] Yang Z, Katsura T, Aihara M, Nakamura M, Nagano K. Investigation into window insulation retrofitting of existing buildings using thin and translucent frame-structure vacuum insulation panels. Energies. 2018;**11**. DOI: 10.3390/en11020298

[8] Timoshenko S, Woinowsky-Krieger S. Theory of Plate and Shell. Columbus, OH, USA: McGraw-Hill Book Company; 1959. pp. 106-108

[9] Young WC, Budynas RG. Roark's Formulas for Stress and Strain. 7th ed. New York, NY, USA: McGraw-Hill Professional; 2001. pp. 508-509

[10] Baetens R, Jelle BP, Thue JV, Tenpierik MJ, Grynning S, Uvsløkk S, et al. Vacuum insulation panels for building applications: A review and beyond. Energy and Buildings. 2010;**42**: 147-172. DOI: 10.1016/j.enbuild.2009.09.005

[11] Arya F, Moss R, Hyde T, Shire S, Henshall P, Eames P. Vacuum enclosures for solar thermal panels. Part 1: Fabrication and hot-box testing. Solar Energy. 2018;**174**:1212-1223. DOI: 10.1016/j.solener.2018.10.064

[12] Arya F, Moss R, Hyde T, Shire S, Henshall P, Eames P. Vacuum enclosures for solar thermal panels. Part 2: Fabrication and hot-box testing. Solar Energy. 2018:1212-1223. DOI: 10.1016/j.solener.2018.10.064

[13] Choi B, Yeo I, Lee J, Kang WK, Song TH. Pillar-supported vacuum insulation panel with multi-layered filler material. International Journal of Heat and Mass Transfer. 2016;**102**:902-910. DOI: 10.1016/j.ijheatmasstransfer.2016.06.032

[14] ANSYS FLUENT Theory Guide; 2011

[15] Kim J, Song TH. Vacuum insulation properties of glass wool and opacified fumed silica under variable pressing

load and vacuum level. International Journal of Heat and Mass Transfer. 2013;**64**:783-791. DOI: 10.1016/j.ijheatmasstransfer.2013.05.012

[16] Fang Y, Eames PC, Norton B, Hyde TJ. Experimental validation of a numerical model for heat transfer in vacuum glazing. Solar Energy. 2006;**80**: 564-577. DOI: 10.1016/j.solener.2005.04.002

[17] Fang Y, Hyde T, Hewitt N, Eames PC, Norton B. Comparison of vacuum glazing thermal performance predicted using two- and three-dimensional models and their experimental validation. Solar Energy Materials & Solar Cells. 2009;**93**: 1492-1498. DOI: 10.1016/j.solmat.2009.03.025

[18] Memon S, Fang Y, Eames PC. The influence of low-temperature surface induction on evacuation, pump-out hole sealing and thermal performance of composite edge-sealed vacuum insulated glazing. Renewable Energy. 2019;**135**:450-464. DOI: 10.1016/j.renene.2018.12.025

[19] Katsura T, Yang Z, Aihara M, Nakamura M, Nagano K. Development of slim and translucent vacuum insulation panels. Journal of Japan Solar Energy Society. 2018;**44**:49-57. DOI: 10.24632/jses.44.2_49

[20] Fang Y, Hyde TJ, Hewitt N. Predicted thermal performance of triple vacuum glazing. Solar Energy. 2010;**84**: 2132-2139. DOI: 10.1016/j.solener.2010.09.002

[21] Palyvos JA. A survey of wind convection coefficient correlations for building envelope energy systems' modeling. Applied Thermal Engineering. 2008;**28**:801-808. DOI: 10.1016/j.applthermaleng.2007.12.005

Chapter 8

Smart Built Environment Including Smart Home, Smart Building and Smart City: Definitions and Applied Technologies

Reyhaneh Karimi, Leila Farahzadi, Samad M.E. Sepasgozar, Sharifeh Sargolzaei, Sanee M. Ebrahimzadeh Sepasgozar, Mohsen Zareian and Akram Nasrolahi

Abstract

Technology, particularly over the past decades, has affected the cities and their components, such as building sectors. Consequently, smart building that has currently utilized various technologies which is incorporated into buildings is the core of the present chapter. It provides a comprehensive overview on smart cities, smart buildings and smart home to address what systems and technologies have been incorporated so far. The aim is to review the smart concepts in built environment with the main focus on smart cities, smart buildings, and smart homes. State-of-the-art and current practices in smart buildings were also reviewed to enlighten a set of directions for future studies. The Chapter is primarily focuses on 51 articles in smart buildings/homes, as per collected from various datasets. It represents a summary of systems utilized and incorporared into smart buildings and homes over the past decade (2010–2020). Additional to different features of smart buildings and homes, is the discussion around various fields and system performances currently utilized in smart buildings/homes. Limitations and future trends and directions is also discussed. In total, such building/home systems were categorized into 6 groups, including: security systems, healthcare systems, energy management systems, building/home management systems, automation systems, and activity/movement recognition systems. Furthermore, there are a number of surveys which investigated the user's acceptance and adoption of the new smart systems in homes and buildings, as presented and summarized thereafter in Tables. The present Chapter is a contribution to a better understanding of the functions and performances of such buildings/homes for further implementation and enhancement so that varying demands of smart citizens are fulfilled and eventually contribute to the development of smart cities.

Keywords: smart cities, smart buildings, smart homes, intelligence, automation, systems function

1. Introduction and research method

Technology over the past decades has significantly influenced cities, including building sectors to a great extend. The term "Smart" was recently coined and developed across all built environment segments - thanks to efforts made to set up highly-functional buildings and constructions. "Smart building" technologies are currently recognized as a promising solution to enhance flexibility and efficiency of built environment. Such buildings are thus recognized as the main core of diffenerent Smart city components i.e., Smart Infrastructure, Smart Healthcare, Smart Energy, Smart Mobility, Smart Technologies, Smart Construction, Smart Governance and Education, Smart Citizens, and Smart Transportation, as illustrated in **Figure 1**.

The present Chapter reviews the concepts of Smart Built Environment with the main focus on smart cities [1], smart buildings, and smart homes. State-of-the-art and current practices made in smart buildings are reviewed, and a set of directions for future studies are suggested. The main question to address is what systems and technologies have been utilized and incorporated in smart buildings/homes so far.

As such, a holistic view of the smart cities are provided. Papers related to Smart building, the existing definitions of Smart and transformations made over the past decades, how it was categorized, contributions made to the building sectors and current research reported in smart buildings are reviewed. Section 3 is focused on smart homes – the most recurrent concentration as the primary focus of smart buildings. How smart homes are defined and emerged are outlined and a timeline of evolvement are depicted.

A large body of literature (51 papers) in smart buildings and smart homes is reviewed, and practical benefits of smart homes are discussed. Articles were

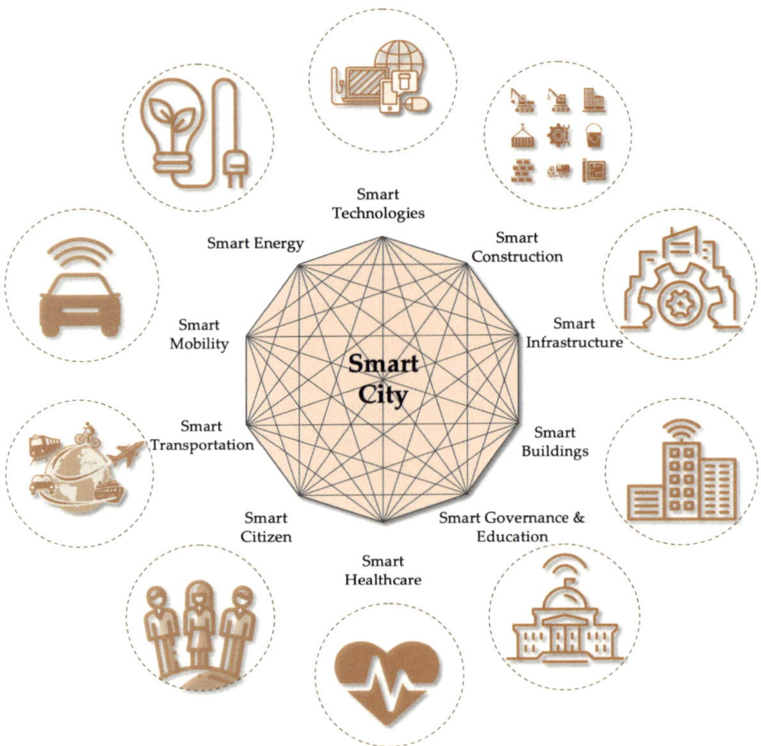

Figure 1.
Smart city components and the connections among them.

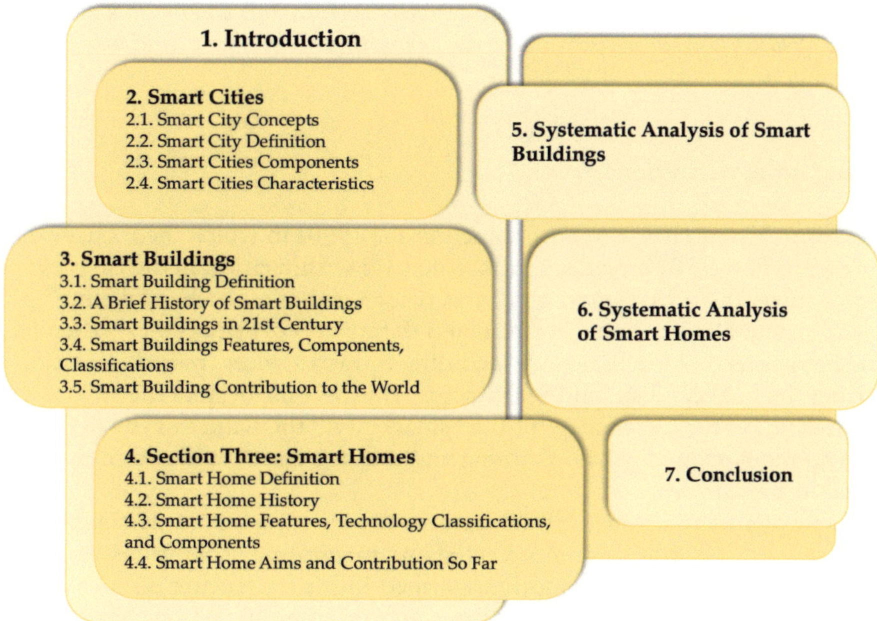

Figure 2.
The framework summarizing the main themes reviewed in this chapter.

extracted from a variety of datasets such as Elsevier, Springer and IEEE with the following keywords searched for relative papers: "smart building" and "automated building" for the part related to smart buildings and "smart home" and "automated home" for the part related to smart homes. as per published between 2010 and 2020. Different technologies utilized in smart homes, features, findings, and limitations were also discussed. Finally, potential research trends in the future advancement of smart home applications and components are outlined.

The most interesting findings of the present chapter are: (1) smart buildings and homes categories (functions): security, healthcare, energy management, building/home management, automation, and activity/movement recognition; (2) Suggestions made for future research can serve as a useful collection for students and researchers and assist with identifing gaps and developing ideas to address future research; (3) Finally, key findings of each article are summarized in Tables, therefore provide readers with key contributions made. **Figure 2** illustrates the framework on which the whole Chapter is organized.

2. Smart cities

Concepts, definitions, dimensions, and characteristics of smart cities are presented as follows.

2.1 Smart cities concepts

Originally traced back to the Smart Growth Movement in the late 1990s, the concept of Smart Cities has recently gained much attention and recognized as an important paradigm in intelligent urban development and sustainable socio-economic growth, [2]. Smart cities in the urban planners' is profoundly debated on the future of cities, the diffusion of smart city initiatives in countries with different requirements and contextual conditions (e.g., developed or developing nations) which makes it difficult to identify common definitions and trends on a global scale. In fact, a globally-agreed definition of the term Smart City or related attributes are still missing.

2.2 Smart cities definition

The term "smart city" was first coined in the 1990s in which the focus was on the impact of new ICT on modern infrastructures within cities. In the first few years of 21st century, the term "smart city" was recognized as an "urban labeling" phenomenon. Generally, in the most presented definitions of "smart city", there are several keywords that are repeated, including infrastructures, resources, efficiency, increasing the quality, data, and technology. This reveals, despite the lack of a unique definition for the concept of the "smart city", the nature of the definitions and the opinions around urban planning and management are in common. As expected in a smart city:

City managers can use city infrastructures and resources in an efficient way. The key feature of the smart city is monitoring and integrating "conditions of all of its critical infrastructures" as previously described [3]. Infrastrucure refers to "roads, bridges, tunnels, rails, subways, airports, seaports, communications, water, power, even major buildings, can better optimize own resources, plan preventive maintenance activities, and monitor security aspects while maximizing services to citizens". As earlier stated [4, 5] "A smart city infuses information into its physical infrastructure to improve conveniences, facilitate mobility, add efficiencies, conserve energy, improve the quality of air and water, identify problems and fix them quickly, recover rapidly from disasters, collect data to make better decisions, deploy resources effectively, and share data to enable collaboration across entities and domains". Another quote [6] also indicated "A smart city means uses all available technology and resources in an intelligent and coordinated manner to develop urban centers that are at once integrated, habitable, and sustainable".

Furthermore, a smart city is anticipated to use information for a dynamism in the economy and resource sustainability purposes. For instance, it was stated [7] that: "Two main streams of research ideas: (1) smart cities should do everything related to governance and economy using new thinking paradigms and (2) smart cities are all about networks of sensors, smart devices, real-time data, and information and communication technologies (ICT) integration in every aspect of human life". Another definition is [8] "a city well performing in a forward-looking way in economy, people, governance, mobility, environment, and living, built on the smart combination of endowments and activities of self-decisive, independent and aware citizens". Smart city generally refers to the search for and identifing intelligent solutions which allow modern cities to enhance the quality of the services provided to citizens". Smart cities are profoundly dependent on big data [1, 9, 10], ICT to provide more efficient services to citizens, to monitor and optimize existing infrastructure, to boost collaborations among different economic actors, and to encourage innovative business models in both private and public sectors". **Figure 3** presents a typical network of a smart city including data server and connections to different buildings and infrastructrues.

The concept of "smart city" aimed to increase the efficiency of urban systems operations in urban environments. In the past century, increased poulations and habitats in cities made services to citizens inefficient. Furthermore, from a sustainability and carbon emission footprints perspective, the excessive use of limited resources has pushed the situation into a more hazardous and unsafe environement [11]. It is therefore of prime importance to seek for tools/plans for optimization

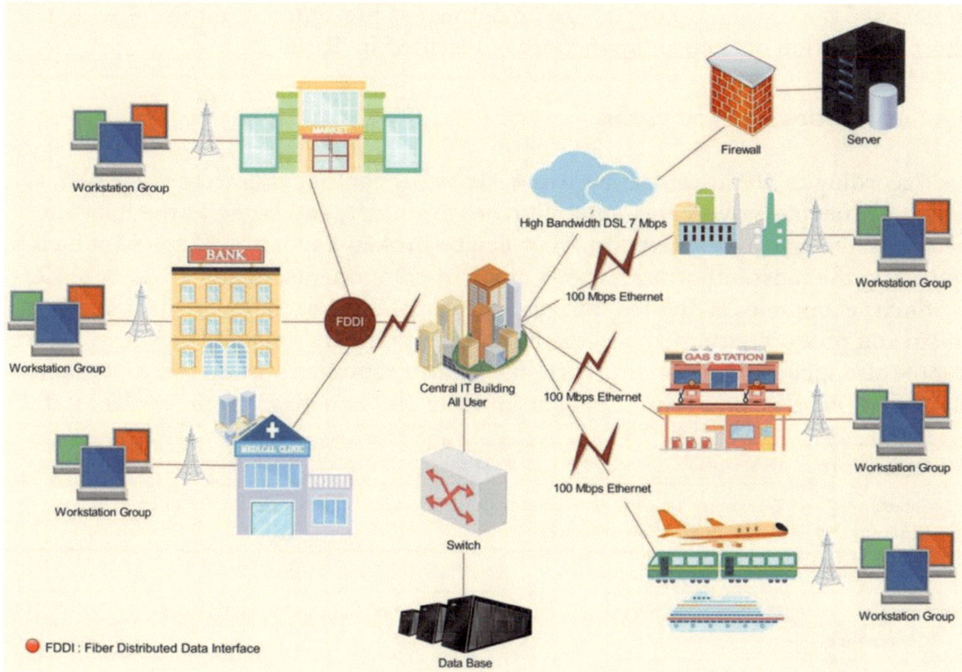

Figure 3.
A typical smart city network.

purposes of urban resources and infrastructures, where data and technologies are gathered in smart city.

2.3 Smart cities components

To clarify what constitutes a smart city, researchers have devided this concept into many features and components, to justify the complexity involved in smart city concept in a holistic way. Attempts made to delineate the features of a smart city

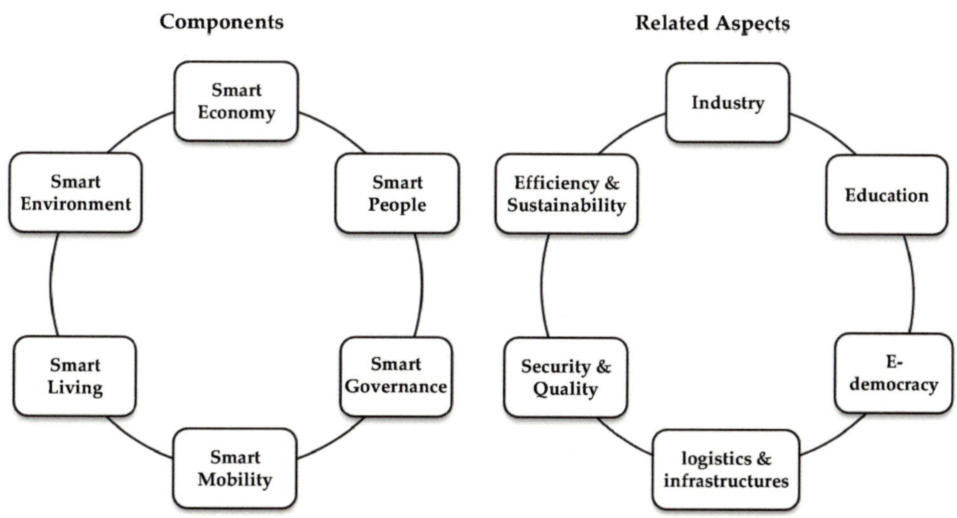

Figure 4.
Key components of a smart city and related aspects.

indicated 6 possible components can be defined as presented in **Figure 4** with a short description of such components as presented in **Table 1**.

2.4 Smart cities characteristics

According to literature and **Figure 4**, six smart components were identified, i.e., economy, people, governance, mobility, environment, and living as the relevant group characterizing a smart city. They can be broken down into 31 relevant factors reflecting the most important aspects of smart components, as shown in **Table 2**.

Smart economics is divided into micro-components, each demonstrating dynamism and productivity, such as creativity, entrepreneurship, and flexibility. Smart people also include components that, like smart economics, imply creativity, and flexibility. Furthermoer, smart people generate and benefit from the social capital

Components	Description
Smart Economy	Eeconomy which is associated with the presence of industries in ICT or employing ICT in production processes.
Smart People	Refered to people possessing creativity, diversity, and education [13]
Smart Governance	Various stakeholders are engaged in decision making and public services.
Smart Mobility	Refers to the use of ICT in modern transport technologies to improve urban traffic.
Smart Living	The smart people factor comprises various aspects, such as affinity to lifelong learning, social and ethnic plurality, flexibility, creativity, cosmopolitanism, open-mindedness, and participation in public life.
Smart Environment	Refers to efficiency and sustainability of environment, while smart technologies are in use.

Table 1.
A brief description of SCs components [12].

Smart Economy	Smart People
• Innovative spirit • Entrepreneurship • Economic image and trademarks • Productivity • Flexibility of labor market • International embeddedness • Ability to transform	• Level of qualification • Affinity to lifelong learning • Social and ethnic plurality • Flexibility • Creativity • Cosmopolitanism/Open-mindedness • Participation in public life
Smart Governance	**Smart Mobility**
• Participation in decision-making • Public and social services • Transparent governance • Political strategies & perspectives	• Local accessibility • (Inter-)national accessibility • Availability of ICT-infrastructure • Sustainable, innovative and safe transport systems
Smart Environment	**Smart Living**
• Lack of pollution of natural conditions • Pollution • Environmental protection • Sustainable resource management	• Cultural facilities • Health conditions • Individual safety • Housing quality • Education facilities • Touristic • Social cohesion

Table 2.
List of characteristics of Smart Cities [14].

of a city, hence, the concept of smart city indicates a combination of education-training, culture-arts, and business-commerce with hybrid social, cultural, and economic enterprises [15]. Smart governance means various stakeholders are engaged in decision making, political strategies and public services, therfore forms Transparent governance. Smart mobility refers to the use of ICT in modern transport technologies to improve urban traffic. Smart environment, reflecting efficiency and sustainability of urban life, includes micro-components such as ambitiousness towards CO_2 emission reduction strategy, efficient use of electricity and water, area in green space, greenhouse gas emission intensity of energy consumption, policies to contain urban sprawl, and proportion of recycled waste. Smart living reflects a city whose citizens have a high quality of life. In such a city micro-components of smart living can be cultural facilities, health conditions, individual safety, housing quality, education facilities, touristic, and social cohesion.

3. Smart building

In this section, smart building definitions, history, and evolution, different classification and contributions are encapsulated. **Figure 5** illustrates an overview of smart buildings in cooperation with a smart city and its components, after which smart buildings shall be introduced.

3.1 Smart building definition

"Smart Building" points out to various technologies incorporated into buildings. There may not be a definite explanation of what turns a building into a smart one.

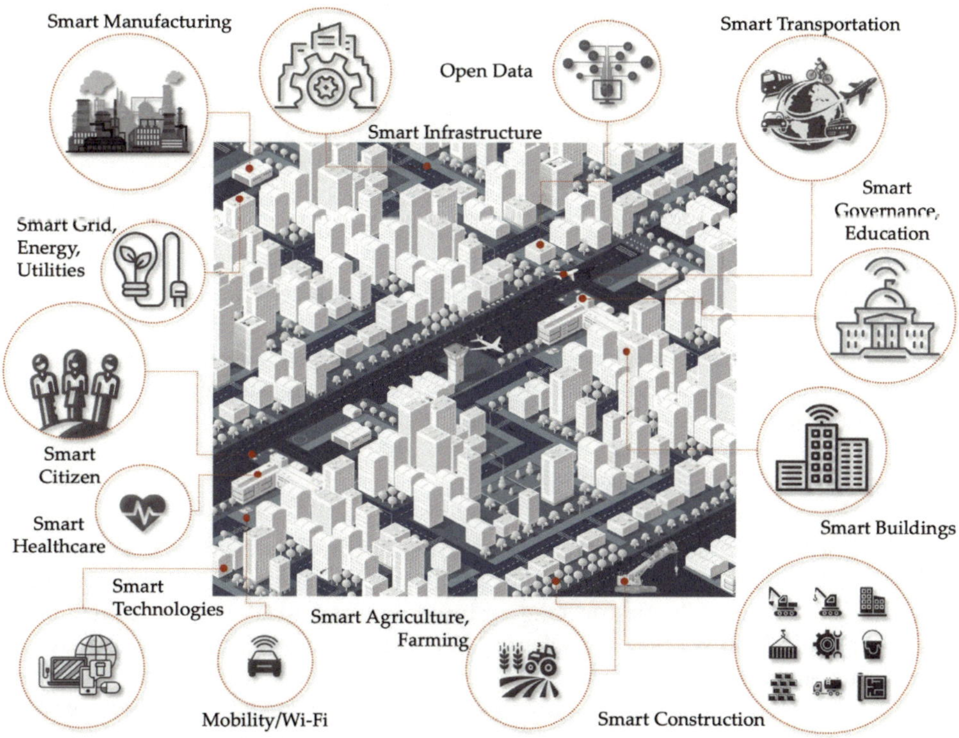

Figure 5.
An overview of smart buildings in addition to other smart components in a smart city.

The term has been defined by various research and it is sometimes interchangeably used as intelligent building or automated building, which are not necessarily the same. That is due to the evolution of the buildings from "conventional", to "automated", to "intelligent", and to "smart" [9, 16].

In 1990, the intelligent building was defined by Powell [17] as a building which controls the environment of itself. Such mode of control did not introduce any interactions made by the user(s). In 1995, Capacity and Institution Building (CIB) proposed the definition of a responsive and dynamic architecture for intelligent buildings which offers the users a cost-efficient, productive and competent environment by 4 fundamental components; i.e., place, process, people and management with interrelations involved [18]. Continental Automated Buildings Association (CABA), in 2002, defined it as a building and the infrastructure which offers occupants an efficient, flexible, safe and comfortable condition by utilizing technological systems, controls and communications, all integrated together [19]. Frank [20] summarized the previously defined explanation of the intelligent building and explained it as an automated building centrally managed to provide users with a safe, comfortable, productive and energy-efficient environment through technological devices which allow sustainability and cost-efficiency. In 2005, Wong et al. [21] discussed that the previously-used definitions centered on minimizing the interaction of users with the building.

It was around 2011 that the term "smart building" started to gain popularity over the previously-used term "intelligent building" [22]. Buckman et al. [23] believe smart buildings are intelligent buildings with additive and combined elements of adjustable control, enterprise, materials and construction. CABA concentrates on the capability of the smart building to adjust itself to different surrounding parameters and conditions using outward context-led data on the behavior of the users [24]. The highlights within the current definition of smart buildings are the concentration on the users and the interaction with the user of a building in high levels [25, 26]. Sinopoli [27] looks at smart buildings holistically and the integration of design, construction and execution where applicable information is acquired within a smart building to facilitate the management of the environment by the users/occupants. In the most recent researches, smart buildings are defined as those which benefit from the efficient and cutting-edge sensors and electronic devices as per connected together and communicate with each other to generate a large amount of data, (refered to as Big Data), to control and manage the building smartly to boost reliability and performance of the building and reduce energy consumption and cost; more importantly to provide a desirable, flexible and safe environment for occupants [28–32]. Smart buildings depict the evolvement of the most up-to-date infrastructure and integration of automated control systems, data computing, and artificial intelligence. The smart building adapts itself to the surroundings according to both inner and outer parameters and can learn [22], interact and self-adapt to the needs of the occupants.

3.2 A brief history of smart buildings

Although smart buildings seem to be a new concept, the origins actually dates back to 1980s, when a new generation of intelligent buildings was constructed in the United States [33]. The first intelligent building was built in Hartfort, the United States in 1984 [20]. It is however believed [34] the concept of intelligent buildings refers to automation rather than intelligence which is rooted back to an earlier time of 1960s, when a building was resourced with an automation system which served as a device for 'work saving time'. In 1973, energy crisis occured which raised awareness on energy consumption style, and subsequently led to the formation of smart features of energy consumption and crime prevention. In the early 1980s, the emergence of two main technologies assisted with the development of the trend; i.e., telecommunications and personal computers. Establishment of networks of technology and real estate enterprises offered a shared telecommunications system to building owners [27]. Integrated single-functional systems for building automation such as access control, security control, lighting control, lift control, electrical systems, fire system, electronic processing and communication of data, text communication and telefax, image communication and TV were generated at single level function as the integration of different subsystems was not possible [34]. In 1990s, technological developments i.e., direct digital control (DDC), video surveillance, access control systems, audio visual systems and cabling systems [27] and integrated multi-functional systems were feasible at building automation system (BAS) and integrated communication system (ICS) level. In 2000s, developments were progressed in the implementation of systems integrated with computers, utilizing Internet protocol (IP) network technologies which boosted network capacity, and remote monitoring and control through the Internet [34]. **Table 3** illustrates the evolution of the smart building, along with time and technology development.

Time	Evolution	The type of building Management
1950s	• Using pneumatic control with compressed air in automated buildings • Execution of fire alarm system [34]	Building Control [36]
1960s	• Emergence of analog and then digital microprocessors • 'work saving time' device implementation	
1970s	• Systems for crime prevention and energy preservation	
1980s	• Construction of the 1st intelligent building • Implementation of single-functional systems • Emergence of telecommunications and personal computers [27]	
1990s	• Implementation of multi-functional systems • Supporting communication among controlled facilities by open protocols	Building Automation
2000s	• Implementation of systems integrated with computers [37] • utilizing Internet protocol (IP) network technologies and boosted network capacity. • Emergence of Wi-Fi • Achieving remote monitoring and control through the Internet	Building Performance
2010s	• Implementation of enterprise network-integrated systems • Delivering multiple services by combining BAS and IT through • Internet Protocol network • Integration and management because of modern IT technologies e.g., Web Services, XML, distance portfolio management and helpdesk management, multimedia communication via cellular phone • Emergence of BIM and embedded intelligence	Smart Building [36]
2020s	• Integration of building and industrial PLC system, cloud computing and IoT, data mining and data-centric energy management, cyber security and protection, hardware and software analytics, the progress of Smart City, and utility meter monitoring and reporting [35]	

Table 3.
The evolution of smart buildings, adapted by the authors from [27, 34–37].

3.3 Smart buildings in the 21st century

Building automation technology and smart buildings have been ever developing since the early 21st century. Smart buildings are now capable of making connections between people, and surrounding environments such that buildings are more responsive to the demands of users and city in a more sustainable fashion which consequently boost human's wellbeing [37, 38]. Currently, the "concept of Industry 4.0" has been put into research with a focus on building sectors. Industry 4.0 is fundamentally the integration of Internet of Things (IoT) and digital data acquired from various sources of sensors and actuators. The Industry 4.0 can potentially enhance safety and substantially lower energy consumption while maintaining comfort in smart buildings [39]. Networking communication technology and inserted micro-electromechanical technologies have recently received much attention. Different devices can be embedded by IoT, which ultimately can enhance sensing capability, identification and communication within the smart systems [40]. Blockchain technology can also be utilized as a measure to control the platform, including the IoT and Building Information Modeling (BIM) [41] in smart buildings [42]. Artificial intelligence (AI)-based systems play an important role for the realization of the five main energy optimization perspectives – users, comfort, safety, design and maintenance in smart buildings [22]. A survey conducted on 45 experts on the future of smart buildings [43] revealed the following technology trends in future smart buildings: (i) Accommodating a single cloud-based IoT platform; (ii) Organizing smart building technology via multiple asset-distinct IoT Platforms. Obviously smart buildings will be more adaptable to the fast change of the demands and needs and focus more on sustainability aspects in the future. Future smart buildings will be highly user-centric, thanks to advanced and innovative analytics, system integration and edge computing as incorporated throughout smart buildings, therefore making remote building management services feasible and boost maintenance.

3.4 Smart buildings features, components, classifications

Major elements of smart buildings can be categorized into four groups; (1) structure; (2) systems; (3) services and (4) interrelations [34]. Another classification includes components of smart buildings into hardware, software, and network based on the concept of IoT and Cloud, data analytics, artificial intelligence, and decentralized energy [44]. New technologies are obviously inevitable in smart buildings and include (but not restricted to) sensor utilization, big data analytics and engineering, fog and cloud computing, development of software engineering, and algorithms of human-computer interaction [45]. A selective classification of smart buildings technology can be compromised of the following components: energy-efficient LED light fixture, occupancy sensors, building dashboards, electrochromic windows or smart glass and indoor positioning [16].

Smart buildings can also be categorized based on the following features: (1) Environmental friendliness – energy and water preservation sustainable design; efficient disposal of wastes; minimizing pollution; (2) flexibility and utilization of space; (3) adding value quality to economic life cycle cost; (4) well-being and health of the occupants; (5) efficient working; (6) security and safety actions – structural failures, fire, earthquake, and disaster; (7) culture– addressing the expectations of the client; (8) innovation and effectiveness of technology; (9) management and construction procedures; (10) sanitation and health [37].

Smart buildings are also defined as three main management solutions: (1) Information system- acquiring data from different devices, collecting, reporting and visualizing the patterns, and enabling decisions for the related actions; (2) Analysis system - further data analysis to find out particular insights where regular utilization consists of modeling to offer measurements and verification abilities e.g., fault detection and diagnostics (FDD); and (3) Control systems - connecting to the assets and controlling assets in the smart building such as heating, ventilation and air conditioning (HVAC) and lighting systems, changing set-points using schedules and adjusting conditions [46].

3.5 Smart building contribution to the world

Smart buildings offer the most effective way for cost efficiency and the utilization of technology systems of the building. During a building design phase, the whole building technology systems design is coordinated by one designer and combined into the construction documents, which indicates each system and meets the elements of the common system or the integration of bases for the systems. It consists of cabling, cable pathways, equipment rooms, databases of system, and devices communications protocols. The final design is referred to as a contractor. Deficiencies in the design and construction process is therefore minimized with time and costs being saved. Given technology systems of the building is integrated on the horizontal and vertical aspects in building operations, therefore it gives occupants access to building's operation data and information and manage respective section. Beneficiary to building owners and developers, it contributes to property value improvement. To facility and property managers, it offers further efficient subsystems and effective management alternatives e.g., system management consolidation. As the design and construction are combined proportionally, it can assist architects, engineers, and construction contractors achieve higher efficiencies and contributes to savings in the project [27].

In addition to management systems, smart buildings can significantly influence other sectors; i.e., retailing (smart shops), health cares (in smart hospitals/homes), and safety and security (intrusion/detection systems) [47].

Smart buildings offer a variety of solutions and benefits e.g., detecting the number and activity of the user, healthcare, energy usage and sustainability-related services i.e., HVAC systems control, heat control, comfort modeling, device control, load predicting, power management, lighting control, longevity and satisfaction [27, 31, 48]. Smart technologies can also offer opportunities for plug loads, HVAC systems, lighting, window shading, human operation, automated system optimization, and connected distributed generation and power [29]. Smart buildings' passive design strategies smartly engineered in smart buildings makes building systems tailor both environmental situations and user needs [32]. Safety, productivity and comfort are provided to the users in smart buildings. Users' comfort is ensured through smart HVAC systems, smart water, gas, electricity metering, occupancy detection and monitoring systems and even hybrid charging technology of vehicles [31]. Contributions of smart building can be summarized as follows: (1) occupants comfort: smart buildings learn the occupants' behavior and make efforts to enhance occupants' comfort level; (2) saving energy: smart buildings can significantly reduce energy usage, and costs associated; (3) saving time: the automation of daily routines can significantly save time; (4) safety: detection of fire, gas leakage, using self-diagnosis systems, capable of alerting, enhance safety level; (5) expert systems: embedded in smart buildings can hold the knowledge of the areas; (6) healthcare: in smart buildings, health-related decisions are of the highest priority in which services e.g., suitable temperature, air condition factors, and light intensity are supplied; (7) assistance: as smart buildings can improve life quality of elderly and disabled people by providing comfortable, safe and supportive environments; For instance, providing assistance in daily routines, alerting the family and/or social services in case of emergency, mitigating loneliness by connecting to other individuals via Internet [44].

4. Smart homes

In this section, smart home definitions, history, and evolution, different classification and contributions are discussed.

4.1 Smart home definition

Smart homes are recognized as an important focal point in smart buildings [45] and described in section one. The term "Smart Home" is defined [49] as a home benefiting from a "Controller" to combine the different automation systems. The most well-known controllers in homes are connected to computers within the programming period only, after which home control tasks are performed independently. Home systems integration leads to better communication among systems via home controller. Consequently, it triggers a single button and voice control of the different systems in the smart home at the same time, in scheduled scenarios or operating modes. In other words [50], smart homes are recognized as automated buildings benefiting from installed detection and control systems, i.e., air conditioning and heating, ventilation, lighting, hardware, and security systems [51]. As new systems with switches and sensors can communicate with a central axis (typically referred to as gateways). The gateways are systems which control user's interfaces and connection to a tablet, cell phone, or computer. Such systems network connection is managed by IoT. A smart home is also defined as a residential extension of building automation which includes the control and automation of the whole utilized technology [52]. As such, a smart home comprises appliances, lighting, heating, air conditioning, televisions, computers, entertainment systems, big home appliances e.g., washing machines, fridges, and security systems (e.g., camera systems with the function to interact with each other), and controlled remotely by means of a schedule, cell phone or internet. Such systems include switches and sensors connected to a central hub, and controlled by a user, a wall-mounted terminal or mobile unit via internet cloud services.

4.2 Smart home history

The idea of smart home was initially formed in the 1990s [53] and a brief timeline of smart home evolution is illustrated in **Figure 6**.

4.3 Smart home features, technology classifications, and components

Smart homes include five basic features; i.e., (1) Automation: the capability of accommodating automatic devices or carrying out automatic functions; (2) Multi-functionality: capability of conducting different tasks or creating various outcomes; (3) Adaptability: capability of learning, predicting and meeting the wants of users; (4) Interactivity: capacity to provide and allow communications among users; (5) Efficiency: capability to perform functions conveniently and save time and costs [53].

The smart home technologies benefit from four classification criteria, including (1) collecting the data: required data is collected after which being accessible; (2) processing the data: the collected data is processed and analyzed and can be

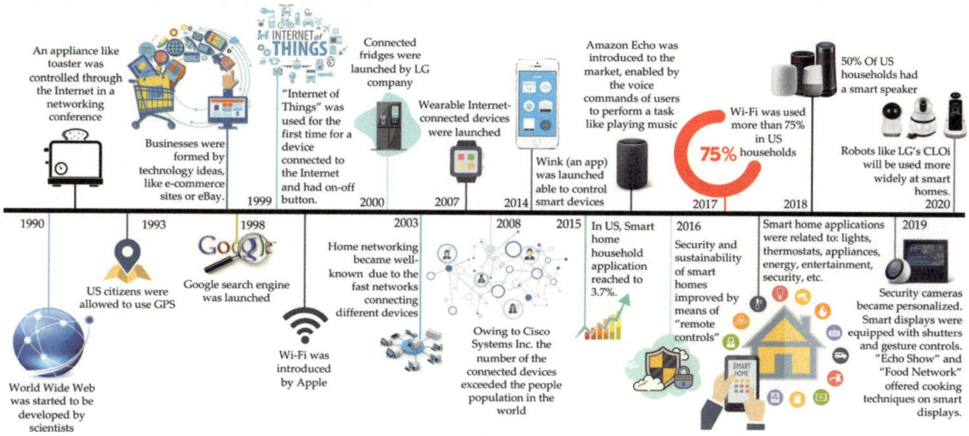

Figure 6.
Smart home evolution timeline.

combined; (3) representing the data: the data is accessible to the users; (4) controlling and interacting abilities: the status is accessible to the users and the functions of related technologies are monitored.

Smart home components can also be categorized [52] as follows: (1) Sensors which collect indoor and outdoor data of home and measure the conditions of the home. Sensors are therefore connected to the home and to the embedded home devices. Data is collected and sent through local networks to smart home servers constantly; (2) Processors which carry out local and combined activities. Processors are also connected to the cloud for extended source-based applications. The data collected by the sensors is consequently processed by processes via local servers; (3) A set of softwares are wrapped as APIs which allows external apps performance and follow the pre-defined parameter format. APIs can process data and manage essential tasks; (4) Actuators which provide and perform commands within the server or other control devices. Actuators can translate activities which are required to the command syntax. The system can launch commands to the suitable device processor; (5) Databases which store processed data as collected from sensors and cloud. Databases can also be used for analyzing, presenting, and visualizing data. The processed data is stored in the database for future usage.

4.4 Smart home aims and contributions

Smart homes were initially aimed at enhancing users' comfort level and making daily life easier, and provide occupants with ambient intelligence, remote home control or home automation systems [53]. The advantages of smart homes integrated with IoT can be summarized [50] as follows; (1) "Energy conservation" in which energy waste is lowered using improved IoT technology, higher quality and reliability of devices, improvements in system security and applications, and sensible use of devices; (2) "Decreased cost of fundamental needs of smart homes" contributing to advancement in smart home technologies that are based on IoT, application of medical technology services, using wireless technology and a security system, using energy conservation devices, and suitable users' behavior; (3) "Healthcare" contributing to medical services, appropriate living for the elderly, easier communication with health centers, treatment alarms, controlling, and monitoring elderly patients; (4) "Entertainment and comfort" with more comfortable, convenient to use and control, and easy to pay bills.

This section comprises Tables of the existing research on smart buildings and smart homes, respectively. **Table 4** summarizes 5 papers related to smart buildings. **Table 5** shows the number of databases, and papers extracted per each database. **Tables 6** and 7 present the summary of reviewing 46 papers related to smart home technologies.

Title, Category & Objective	The proposed system function and details	Finding or Contribution Limitation & Suggestions
• Modeling, Analyzing and Predicting Security Cascading Attacks in Smart Buildings Systems-of-Systems (SoS) [54] • To predict attacks at architecture stage using software engineering approaches	• A Model-Driven Engineering method, Systems-of-Systems Security to benefit from: (1) a modeling language (SoSSecML) for secure SoS modeling (2) Multi-Agent Systems (MAS) for security analysis of SoS architectures. • A case study conducted on a real smart building SoS.	• Cascading attacks i.e., individual attacks, Service Denial were detected. • SoSSec to be extended to address various strange behaviors and non-functional properties i.e., safety and trust.
• Prediction Model for Personal Thermal Comfort in Naturally Ventilated Smart Buildings [55] • To enhance non-air-conditioned or naturally ventilated buildings to increase users' satisfaction	• Different datasets were used from various climates and seasons to identify an optimum feature set. Variables were selected and processed by supervised machine learning techniques i.e., Support Vector Machines (SVM) and Naïve Bayes Classifier.	• Indoor and outdoor temperature and humidity, under any given climate or season, were identified.
• An IoT-Based Thermal Model Learning Framework [48] • To propose a learning platform of plug and play to recognize the thermal model of each smart building zone based on low-resolution data from intelligent thermostats without human intervention	• To automatically train the thermal properties of the building directly from historical data of building operation. Inputs included: temperatures, relative humidity, time of the day, day of the week, and seasons. • Thermal model learning pipeline includes: (1) data acquisition from thermostats; (2) thermal model learning on cloud infrastructure or an edge device; (3) ultimate learnt model provision to other applications of smart control. • Five common supervised models for learning tested include: polynomial regression (PR), support vector regression (SVR), random forest (RF), extreme gradient boost (XGB), and neural network (NN). The model with the best function is applied to train the prediction model of the temperature gradient. • An office building on a real-based data collection was conducted for the developed thermal model learning platform.	• A trustworthy thermal comfort assessment was delivered when operating intelligent control. • To be incorporated with an IoT-led BEM and to deliver a precise thermal model to other intelligent air conditioning control algorithms with no extra hardware cost. – the process of learning could provide learnt model on cloud infrastructure and edge device, for a single thermal zone, in minutes. • Does not need configuration for each building, thus enhances the speed to adopting this technology in the market. • Future study: (1) platform assessment on other types of buildings with more changeable occupancy and operation behavior; (2) to adopt in multi-phase cooling/heating systems.

Title, Category & Objective	The proposed system function and details	Finding or Contribution Limitation & Suggestions
• Application of Integrated Building Information Modeling, IoT and Blockchain Technologies in System Design of a Smart Building [42] • To utilize blockchain technology as a securing and controlling framework incorporated with IoT and BIM.	• Effective energy usage and environmental trace control. Blockchain provides an extra layer of data security. The suggested system design allows amending and updating data in cases of reconstruction or renovations; it prohibits unauthorized users from acquiring data on the operation of a building, i.e., the position of sensors and cameras. Data can be acquired from the IoT systems and recorded in Blockchain prior to storage.	• Secure maintenance and management of data concerning. • Can be used in many types of public and private buildings. • Integrated BIM, IoT and blockchain technologies make a novel framework assisting digital transformation in the AEC.
• An Intelligent and Smart Environment Monitoring System for Healthcare [56] • To propose a straightforward android-based wound healing response to monitor external environmental factors for wound recovery purposes.	• Temperature, humidity, smoke, and dust are monitored by microcontroller with multi-sensors. • It consists of (1) Arduino Sensors-based data collection, (2) data recording, (3) data assessment by NN, (4) use of Android App to indicate wound healing level. • It was tested in a house, a park and an industrial building.	• Prediction accuracy rate ranged 96–99%. • Future studies include: (1) applying other machine learning methods i.e., Fuzzy system, Decision Trees, and KNN; (2) using more sensors for wound microenvironment monitoring;

Table 4.
Reviewed papers related to smart buildings.

Data bases	Papers reviewed
Elsevier	16
Springer	16
IEEE	14
Total	**46**

Table 5.
Detailed numbers of databases, and papers extracted per each database (reference: Authors).

Title, Category & Objective	The proposed system function and details	Finding or Contribution Limitation & Suggestions
• Activity-Aware Sensor Cycling for Human Activity Monitoring [57] • To predict the residents' behavior accurately to make a balance in "energy consumption".	• An activity-aware sensor cycling (ASC) to manage uncommon and unpredicted activities and makes a balance between sensors' energy usage by a scheduling algorithm.	• ASC sensor detects 99% of activities and warranties the network lifetime for 2000 h.
• Enabling IoT for In-Home Rehabilitation: Accelerometer Signals Classification Methods for Activity and Movement Recognition [58]	• The system is using an accelerometer sensor on patient's smartphone platform. Patient's movement is recognized upon physical	• SVM-based accuracy is >90% in activity recognition and > 99% in movement recognition. • Future works: A combination of

Title, Category & Objective	The proposed system function and details	Finding or Contribution Limitation & Suggestions
• A "healthcare system" aimed at home treatment and therapy using an "activity/movement recognition system".	activities using support vector machines (SVMs), decision trees, and dynamic time warping (DTW).	SVM and DTW as SVM alone was unable to recognize movement recognition in few cases.
• An Intelligent Human Behavior-Based Reasoning Model for Service Prediction [59] • "Energy management system" aimed to decrease energy waste using a service prediction model and a smart home platform.	• A human behavior-based reasoning (HBR) based on auto regressive prophecy and case-based reasoning (CBR) was used.	• Better performance and higher prediction accuracy than other existing models, particularly in the peak of activities. • A combination of linear prediction and human behavior reasoning contributed to saving energy. Also a combination of embedded systems, ZigBee wireless communications, and appliance control protocols offered less problems when controlling smart homes.

Table 6.
Reviewed papers related to smart homes.

Title & Journal Category & Objective	Details of the proposed survey	Finding or Contribution Limitation & Suggestions
• Patients' Adoption of WSN-Based Healthcare Systems: An Integrated Model of Facilitators and Barriers [60] • This "healthcare system" aimed to find factors which may influence individual's intention to use wireless-sensor-network in smart home healthcare system. Based on which the developers can be more aware of key factors in designing such systems to encourage people to use it.	• A combination of qualitative and quantitative methods is used. • Factors which affect system adoption were identified, (2) A model, by which a number of hypotheses on adoption factors can be proved was proposed. • A semi structured interview of 6 home healthcare patients and 9 medical experts, plus a survey questionnaire from 140 respondents.	• The encouraging factors include ability to enhance the life quality and supporting emotionally. While discouraging factors are their negative feeling of privacy, attachment, and costs. • Future work: to enlarge sample size, broaden culture, age and gender range.
• Benefits and Risks of Smart Home Technologies [61] • The paper focused on the "attitude" towards smart home and investigated the understanding of 3 groups of users.	• A comparative analysis on the ideas and understanding of 3 groups of current users (n = 42), future users (n = 1025) and industry (n = 62) was conducted on the pros and cons of SHS and the policymaker's roles. -The paper benefits from field studies (survey), content analysis, and statistical analysis	• (1) Future users have positive perception on different usage of SHTs i.e., energy management, indoor controlling and promoting safety. (2) Available users and future users both expressed concerns on transferring self-reliance, independence and data security; (3) Policymakers positively influence in SHT market by formulating standards/guidelines.
• The Impact of Social Connectedness and Interaction Types on Perceived Social Support and Companionship [62] • Focused on "human-device interaction and social adoption" to investigate the notion of	• A human–device social connectedness survey where human–device interaction was classified into two categories (1) internal relation (Inner Social Connectedness-ISC) defined as relation between user - devices in user's own smart home; (2) external relation. 2 types of engagements:	• (1) both type of human-device relations boost user's social experience; (2) Direct engagement with internal relation is more efficient; • Future studies: building up more objective factors rather than subjective ones.

Title & Journal Category & Objective	Details of the proposed survey	Finding or Contribution Limitation & Suggestions
relation and engagement between user and smart devices.	direct/unmediated (user is engaged with the device directly) and indirect/ mediated (user is engaged with the device via an agent). • Controlled experimental study in a smart home prototype using questionnaire with 96 participants, Wizard-of-Oz" technique, AXURE RP 8.0 prototype tool, and statistical analysis.	
• Comprehensive Approaches to User Acceptance of IoT in a Smart Home Environment [63] • To identify main criteria of user's approval towards IoT technologies in smart homes and explore a thorough model containing five possible factors and technology acceptance model (TAM).	• A survey (interviews and questionnaire) on user's acceptance of IoT. Citeria included: enjoyment, compatibility, connectedness, ease of use, usefulness, attitude, intention to use, control, cost collected from 1057 participants. • Confirmatory Factor Analysis (CFA) Structural Equation Modeling (SEM) and Statistical analysis were used.	• Compatibility, connectedness, control, and cost are key factors of technology approval from user's perspectives. • Future research: promoting IoT technologies in a smart home by paying attention to the user experience.
• Convenience and Energy Consumption in the Smart Home of the Future: Industry Visions from Australia and Beyond [64] • The paper focused on the "industry vision" towards energy management security systems" to investigate the vision on the convenient life notion of smart homes and energy savings implications.	• Survey of industry visions on the convenience and energy consumption. Analysis of online articles, magazines and in-depth interviews with 10 industry experts in smart homes were carried out.	• The conflict of simplicity and complications among the vision of smart home industry. • Future research: how to evade and interrupt the notion of convenience in Smart home.

Table 7.
Reviewed tables related to people acceptance of smart home technologies.

5. Systematic analysis of smart buildings

The smart building literature summarized in **Table 4** included research of the latest studies. Selection of papers were made based on a varying spectrum of smart buildings deemed to be of prime importance so that a holistic overview of technologies embedded in smart buildings are provided. Examples included security systems, [54, 55] users' thermal comfort [48], blockchain technology [42] and external environment monitoring system in smart buildings [56].

6. Systematic analysis of smart homes

The papers reviewed in this section were found using a combination of the following keywords: "smart home" and "automated home" as per published on Elsevier, Springer and IEEE databases. Databases as stated above are well-recognized as the most highly-cited ones than other databases in the field. All papers were published between 2010 - 2019 reporting various technologies utilized in smart homes, as well as surveys and questionaries which reported public viewpoints on technologies as such.

From the above 46 articles in smart homes, 2 review Tables are formed and presented. **Table 6** focuses on the technological systems according to different functions and performances in smart homes; **Table 7** presents a review of surveys investigating the people's acceptance or adoption of newly proposed smart homes-based technologies. Three articles as examples are shown in **Table 6** as following and the rest are included in the appendix.

Articles reviewed in the **Tables 6** and **8** (appendix) are categorized in different clusters according to function and objectives. Articles [57, 80, 87, 100] were focused on the "activity/movement recognition systems"; articles [60, 65, 68–72, 92, 97] were focused on the "healthcare systems"; articles [59, 67, 74, 75, 79, 81, 82, 84–86, 89, 93, 95, 98] were focused on the "energy management systems"; articles [66, 78, 88, 90, 91, 94] were focused on the "security systems"; article [73, 96] focusing on the "home management systems"; article [99] was focused on the "automation systems; Also articles from a combination of systems in a frameworks such as [58] focused on the "activity/movement recognition systems and healthcare systems"; articles [101, 102] were focused on the "home management systems and automation systems"; article [83] was focused on the "energy management systems and security systems"; and article [75] was focused on the "energy management and home management systems".

Table 7 presents the papers that focused on the survey from user or market to evaluate their attitude and vision, acceptance or adoption of the smart home technologies.

7. Conclusion

Ambitious projects using modern technologies to develop the city are increasingly expanding worldwide. With such ever increasing interest, authorities, officials and managers are therefore strongly pursuing the idea of incorporating varied technologies in different buildings. Given rapid growth of technology embedded in the different aspects of buildings and the increasing trends in smart cities, smart buildings, and smart homes, the present chapter reviewed and summarized existing technologies/systems currently adopted in buildings/homes. A general overview of smart city components, smart building functions and clusters, . A holistic view on smart cities/buildings/homes, history and evolution, advances made over the past decade, components, classifications, current gaps, limitations, and future trends were provided. Papers in smart buildings and smart homes were searched from reliable datasets i.e., Elsevier, Springer, and IEEE, reviewed and summarized in Tables. Papers were categorized according to the proposed systems and frameworks in the following clusters: security systems, healthcare systems, energy management systems, building/home management systems, automation systems, and activity/movement recognition systems.

Overall, the positive features, or the expectations of the smart buildings, homes and cities can be summarized as follows: higher energy efficiency, connectedness, compatibility, energy management and controlling, promoting safety, enhancing life quality and provide emotional support to users' and cost efficiency.

Despite all such positive features, there exist undesirable aspects that should be taken into account to achieve the smartness, e.g., pitfals oriented around transferring self-reliance, independence, data security, privacy, and costs need to be appropriately addressed.

Attempts, however, have been constantly made to eliminate such negative perspectives. Examples include efforts made to reduce negative perceptions of the public via mechanisms i.e., guaranteeing the reliability and simple use of smart technologies where policymakers and expert panels can facilitate the expansion of smart technology market by enforcing/developing standards and guidelines. Suggesed papers for future works are listed as follows:

- To increase mobile battery capacity to support wearable sensors [68].
- To improve the reasoning process, evaluate ethical, clinical and economical aspects further to the functionality and usability aspects. [71].
- To expand the number of behavior situations [92].
- To include further tasks based on different scenarios and develop application platforms such as MicroApp [101].
- To promote IoT technologies in a smart home using users' experiences [63].
- To extend platform application on other types of buildings and adopt in multi-phase cooling/heating systems [48].

Appendix

43 articles are summarized in **Table 8** as an addition to the **Table 6**.

Title, Category & Objective	The proposed system function and details	Finding or Contribution Limitation & Suggestions
• Fog Assisted-IoT Enabled Patient Health Monitoring [65] • A "healthcare system" to monitor the patients remotely, classify events based on fog computing, mine the patients' health data at cloud layer, and make real-time decisions, based on fog computing at smart gateway.	• Devices implemented in the data acquisition layer included: bio sensors, smart wearables, smartphone, smart devices, RFIDs, and smart monitors. Recognition of patients' events and choosing between normal and abnormal events was based on the Bayesian belief network (BBN) by embedded data mining, distributed storage, and notification services at the edge of the network, and triggering-based data transmission.	• Model validity was assessed using 67 patients health data living in smart homes for a month. • Significantly accurate and responsive model in recognition of the event state compared to other algorithms.
• Smart Secure Homes: A Survey of Smart Home Technologies that Sense, Assess, and Respond to Security Threats [66] • Eisting technologies in "security system" were reviewed.	• (1) Identified a varying spectrum of security problems (2) Summarized existing independent sensors and methods to detect threats.	• Future work: new security risks to be investigated. Unusual circumstances to be automated using a combination of sensors to sense incongruity more accurately and faster.
• A Generic User Interface for Energy Management in Smart Homes [67] • Aimed to introduce a generic user interface for operation systems, which can be easily adapted to different types of buildings to aid "energy management systems".	• (1) Various roles and permissions were introduced. (2) A number of data models were designed. (3) Few functional components are being known to support different kinds of issues of smart homes. • Design, function, and use of the	• The system can fullfill remote reachability, responsiveness, role management, flexibility and generality. It does not meet the system configuration. It proved to be >90% functional, and informative on the amount of energy used in homes and contributed to save costs.

Title, Category & Objective	The proposed system function and details	Finding or Contribution Limitation & Suggestions
	system is firstly evaluated qualitatively by a prototype after which quantitatively assessed by questionnaire as distributed to the system users.	• Future work: (1) The configuration issue to be solved, therefore extra interface is required. To collect and analyze various buildings configuration requirement. (2) Integrating other features e.g., security or music.
• A Health Gateway for Mobile Monitoring in Nursing Home [68] • The "healthcare system aimed to monitor people, particularly in nursing homes, remotely.	• The proposed remote health monitoring system (RHMS) operates with wearable devices in nursing homes, public healthcare centers, and smart homes in which elderly or patients live. The gateway consists of six components: MCU, camera with Wi-Fi, RFID reader, infrared temperature module, Wi-Fi module to convey health data, and Bluetooth to receive data from wearable sensors. Identity, vital signs, temperature, location, and falling action detection was automatically collected from each individual.	• The gateway conveys gathered data through Wi-Fi to the server. In case data were not normal, the gateway sends an alarm to the serve. The server sends out patient's location to physicians' mobiles to find patient. • Future work: (1) Investigating the quality of communication network which transfers the data (2) Increasing mobile battery capacity, which supports wearable sensors.
• A Hybrid Key Item Locating Method to Assist Elderly Daily Life Using IoT [69] • A "healthcare system" aimed to support the elderly living in their homes by storing elderly movements data and provide assistance to easily find necessary items	• (1) Data is collected by sensors and transmitted by Bluetooth, (2) Data is processed by Kernel program, (3) after which findings are visualized by mobile apps. • Sensors are placed in two different positions: in elderly's important items they may miss or fixed in a location in the home. Each item position is shown on mobile phone. • The algorithms combine xBeacon sensing equipment, received signal strength indication positioning, event analysis method, and intelligent cutting algorithm.	• The system can contribute to finding the location of elderly important items and tracking routine behavioral patterns; opportunity for elderly to live without others help.
• A proposal based on BCI system to aid patients with amyotrophic lateral sclerosis [70] • A "healthcare system" to assisst patients with amyotrophic lateral sclerosis (ALS).	• The proposed system (BCI) works with a digital assistant BrainWave sensor called EEG using an android-based operated software to enable ALS patients perform activities such as: controlling TV by Infrared, controlling lighting by SSR28 module, and using an air conditioner. • The BrainWave sensor in the shape of a headset can capture patients' brain frequency and	• 80% of participants' approval in which patients accompolished daily tasks. • Patients' independence and self-esteem was improved. • Future work: developing a prototype in real homes with real patients.

Title, Category & Objective	The proposed system function and details	Finding or Contribution Limitation & Suggestions
	can recognize patients' eye blink strength	
• A Tailored Smart Home for Dementia Care [71] • A "healthcare system" to recognize special needs of patients suffering from dementia and to design a smart home to suit dementia patients.	• Dementia care requirements were identified and collected data were analyzed, after which a prototype was developed. • The prototype consists of a combination of sensors, an android tablet for a digital frame, and a digital wireless radio for evaluating the sound capacity of the smart home, Raspberry Pi (model B) for the gateway, communication interfaces, middleware, managing and monitoring interfaces, reasoning engine, and rule management interface.	Suitable according to stakeholders' opinion with potential application in real smart homes which assisted caregivers. Future work: to (1) evaluate the prototype by more participants, (2) evaluate the prototype in real smart homes, (3) improve the performance of the reasoning process, (4) evaluate ethical, clinical and economic aspects further to suit functionality and usability.
• Aging in Place – How Smart Home Technologies, the Built Environment and Caregiving Intersect [72] • The "healthcare model" aimed to give the elderly the ability to be independent, safe, and maintain well-being in homes.	• (1) Functionality of an automatic light sensor for a potential remote control purposes was evaluated using a 90-year-old widowed woman case study who was living alone (2) Functionality of a video doorbell lock was evaluated using a 79-year-old woman case study who was living alone and had difficulty to open a front door (3) Functionality of the robotic lawn mower was examined using an 82-year-old man living with his wife who was not able to do household tasks any longer.	• Case studies' feedbacks: The light sensor gave Julie the confidence of moving from place to place, and her daughter believes Julie's home is safer. Mixed feelings with the video doorbell lock. Her nephew has a remote-control app on his cellphone to screen the front door happenings. Husband is worried about the maintenance costs and possibility for the lawn mower to be stolen. • Future work: Further studies to be conducted to find ways to entice the elderly to use new technologies.
• An Assistive VLC Technology Using EOG [73] • The proposed system is based upon the "controlling (management) system and healthcare system" to control the devices employed in smart homes and to help the seniors or disabled citizens.	• Eyeglasses equipped with electrooculogram signal based assistive visible light communication is the innovative proposed system. • Surface electrodes made of silver chloride are located around the person's eyes to capture the vertical and horizontal eye movement signals. Signals are processed and conveyed by white light LED and a photodetector. The more eye patterns (e.g., up-down-up pattern or Eye blink) included, the more devices can be controlled. • A digital door lock is employed to test the validation of the proposed scheme.	• A reliable and accurate system contributing to green and safe smart homes assisting seniors or the disabled. • Various eye patterns can be utilized, thus more devices can potentially be controlled: potential contribution to other research areas.

Title, Category & Objective	The proposed system function and details	Finding or Contribution Limitation & Suggestions
• An Open Source Smart Home Management System Based on IoT [74] • The proposed "energy management system" is aimed to solve energy conservation problems.	• A combination of sensors, microcontrollers, devices (i.e., fans and lights), web server and database, software to aid HTTP communication, a PC framework, and mobile app. • The user chooses app model after which, the sensors may turn a device on/off based on evaluations made on room status. Collected data is sent to the server by microcontroller. Upon filtering, data is stored and visualized.	• One-month data collection of room energy usage decreased ~45. • Compared to other systems, the proposed system is reliable, scalable, open source and effective in terms of costs and energy.
• Deep Learning Model for Home Automation and Energy Reduction in Environment Platform [75] • The "energy management and Home management systems" is aimed to propose (1) a novel platform for a smart home environment, which is powerful, interoperable secure and low cost, and (2) a novel algorithm for energy efficiency based on Deep Neural Networks (DNNs).	• A modular platform called SHE and consists of: on-premises deployed agent that works with Z-Wave and ZigBee, a remote control that is mobile, and a cloud service for data processing. • Includes 14 various sensors and actuators i.e., multi-sensor (for detecting temperature, movements, luminance, and humidity); Smart plug (turning the connected devices on/off and recording energy consumption); Window or door sensor; Z-Wave (controlling sensors and actuators); Meters (measuring energy consumption and rain); Dimmer (controlling light); Switch (controlling the garden watering); infrared controller.	• A platform to enhance the interoperability between sensors and actuators was proposed with focus on reducing energy consumption., Two various DNN models (NILM and ELF) were proposed to identify rare energy usage patterns, and suggest users how to reduce energy consumption. • Future work: (1) improving ELF algorithm to reduce false-positive results (2) proposing a model to cover further home appliances.
• Empowering the selection of demand response methods [76] • The "energy management system" framework assists users to identify the best demand response methods and reduce the costs.	• Firstly, the researches gathered user-specific set of Requirements (REQs) by literature analysis and expert interviews. Secondly, SOCs for responding to each REQ is presented by research and expert interviews. Thirdly, Deming's benchmarking is used and matched with each (Sate Of Charge) SOC. All information flow of the framework was analyzed and events were also added to the scenarios. Fourthly, a prototype was implemented to test the appropriateness and practicability of the proposed framework. • A simulation study based on seven types of households was performed to test the framework.	• When the best demand response methods is selected, the users will pay less money, flatten the load profile, and reduce the peak load. • Limitation: evaluation is based upon simulation; in-reality implementation is missing. • Future work: developing other contexts beyond residential buildings.

Title, Category & Objective	The proposed system function and details	Finding or Contribution Limitation & Suggestions
• Fall Detection Using ultra-wide band (UWB) Sensors and Unsupervised Change Detection [77] • The proposed "healthcare system" aimed to detect fall using an unsupervised algorithm.	• The system includes non-wearable ultra-wideband and radar sensors that consist of transmitter and receiver. • The transmitter used in the device spreads the pulses to the environment and the receiver receives and records them. The received pulses are analyzed (by SEParation (SEP) algorithm) to diagnose the problem in case a fall has happened. • The sensor is located over the door frame to detect the movements. Raw data from sensor is sent to the computer.	• 97% accuratcy rate with SEP algorithm • Best choice to detect unsupervised falls and non-falls activities using SEP algorithm. • Supervised algorithm can detect all falls, however, number of labeled samples should reach a threshold to perform as good as unsupervised one.
• Home Automation- An IoT-Based System to Open Security Gates Using Number Plate Recognition and Artificial Neural Networks [78] • The "security system" is proposed an automatic system that opens security gates.	• An ambient light sensor and control circuit are employed to control the environment light. • Vehicles with license plate recognition can work with the system. Image processing is utilized to recognize the license plate and the characters. For optical character recognition, an artificial neural network is utilized.	• 88% and 93% accuracy in training data and license plate character, respectively. • Can be controlled by a mobile app or web interface and generate security gate notifications on mobile.
• Intelligent Energy Efficiency Model Using Artificial TensorFlow Engine [79] • The "energy management system" by this article proposes a framework to help the IoT devices work efficiently with together.	• Three proposed models are: (1) Intelligence awareness target (IAT) that works based on IAT sensor, IAT smart phone, and IAT smart appliance, (2) Intelligence energy efficiency (IE^2S), (3) Intelligence service TAS (IST). • IAT can recognize the data value based on the situation with responsibility to gather the environmental data by a combination of sensors. Data processing and analyzing roles are played by IE^2S as a server and recognizes the users' consumption patterns to enhance the service for working automatically. And the role of predicting, managing, and controlling the service is played by IST.	• Model can can minimize activities of the network deemed to be unnecessary. • Future work: (1) Developing intelligent algorithms with enhanced prediction accuracy, where network can be directly analyzed.
• Latent Feature Learning for Activity Recognition Using Simple Sensors [80] • The "activity recognition system" aimed to recognize activities using deep learning methods.	• A deep and hierarchical autoencoder, which works with different environmental sensors: motion detector sensors, contact switch sensors, pressure mats, mercury contacts, float sensors, and etc. The stacked autoencoders that	• Deep learning outperformed other methods. • Future work: (1) developing convolutional neural networks and recurrent neural networks for activity recognition (2) developing multi-resident activity recognition.

Title, Category & Objective	The proposed system function and details	Finding or Contribution Limitation & Suggestions
	are used in this study are: one-layer denoising autoencoder called DAE and two-layer stacked denoising autoencoder called SDAE. • 3 datasets (3 smart homes) were used for model evaluation purposes.	
• Multi-Objective Power Scheduling Problem in Smart Homes Using Gray Wolf Optimizer [81] • The "energy management algorithm" tends to solve power scheduling problems to reduce energy bills and peak to average ratio and raise users' comfort level.	• A multi-objective gray wolf optimizer, which is an algorithm called GWO.	• The algorithm outperformed genetic as well as other algorithms.
• An Innovative Heuristic Algorithm for IoT-enabled Smart Homes for Developing Countries [82] • The proposed system is for "energy management system" aimed to investigate an improved algorithm for DRSREOD- demand response (DR) combined with renewable energy sources (RESs) and energy storage system optimal dispatch in home energy system managements (HEMs)	• The proposed system uses heuristic algorithm and joins demand response (DR) with optimal dispatch dependent on overabundance existing PV energy, maximum charge/ discharge rates and vector of states of charge (SOC).	• Variety and programming pliability for customers was achieved once MS and multi-objective genetic algorithm/ pareto optimization (MOGA/ PO) was applied to. • Future study includes changes in the behavior of CE, TBD and Pgsize parameters and HEMS estimation time.
• A Novel Smart Energy Theft System (SETS) for IoT-based Smart Home [83] • The proposed "energy management and security systems" aims to introduce a smart system to discover energy stealing activities by aggressors of households' energy systems and warn the customers.	• The proposed smart energy theft system (SETS) uses energy usage data, machine learning models (i.e., MLP, RNN, LSTM, and GRU) and statistical models. • SETS can meter the energy usage of household's devices in real-time manner and compare the real data to find any abnormality.	• This system can boost the safety of IoT-based smart home energy systems to 99.96%.
• A Secured Smart Home Switching System Based on Wireless Communications and Self-energy Harvesting [84] • The proposed "energy management system" aims to employ a safe smart home switching system based on wireless communications and inner energy repository.	• The proposed secured smart home switching system is composed of (1) photovoltaic for storing energy purposes, (2) access control system, (3) smart hub design system. The system unifies electricity access control of a building' energy storing system, and wireless communication for smart switches and sockets.	• A promoted system for safety and energy efficiency in buildings is achieved with: (1) certain users' authority to make the electricity on/off using code; (2) storing energy for active devices using PV systems and wireless communication for smart switches and sockets.
• Load Profile-based Coordination of Appliances in Smart Home [85]	• A profile-matched time-shift (PMS) optimization algorithm approach which diagnoses a	• (1) Lower cost and more energy saving compared to other methods. (2) flexible and

Title, Category & Objective	The proposed system function and details	Finding or Contribution Limitation & Suggestions
• The proposed system is for "energy management system" aimed to study the load pattern of different devices at home and their optimal simultaneous operations to reduce energy consumption and the overall energy price.	series of begin times of various home devices based on the energy usage costs, delay tolerances, and energy sources limits.	can work in various scenarios. • This method can be further applied and examined for various energy sources in a home.
• Online Energy Management for a Sustainable Smart Home with an HVAC Load and Random Occupancy [86] • The proposed system is for "energy management system" aimed to reduce total energy cost of home devices and thermal discomfort cost related to users (with HVAC load and occupancy).	• The algorithm is to make queues related to indoor temperature, electric vehicle charging, and energy storage; and to reduce cost of average time energy use of HVAC. The system works without predicting any system parameters and knowing the HVAC power demand.	• Substantial simulations that pursue real-life trends show the efficacy of the proposed framework. • Future studies: developing online HVAC control in other types of buildings such as commercial buildings.
• A Knowledge-driven Approach for Activity Recognition in Smart Homes Based on Activity Profiling [87] • The proposed "activity recognition and movement system" aims to present an activity recognition platform in smart homes which is able to recognize targeted activities from off the subject data with higher success rate.	• The proposed activity recognition platform consists of: (1) pre-processing data, (2) automatic segmentation, (3) Activity profiling, (4) removing irrelevant information, (5) creating vectors. • It benefits from classification algorithms (Naïve Bayes, SVM and J48) and ANOVA statistical testing.	• Significant improvements made with training data sets in terms of a number of attributes, activities, and the way the sensors are distributed in homes. • The technique recognized the targeted activity from the irrelevant ones with 42% lower inaccuracy.
• A Matrix-based Cross-layer Key Establishment Protocol for Smart Homes [88] • The proposed "security system" aims to generate secret session keys for security between appliances made by different manufacturers in smart homes.	• A matrix-based cross-layer key establishment protocol, with four stages: (1) Home gateway makes system parameters; (2) Home appliances draw out master keys (shared with the home gateway) at physical layer; (3) The home gateway deals out secret key seeds for home appliances; (4) Any two appliances can straightly make a secret session key at higher layers.	• (1) Any two appliances can directly establish a secret session key without using any pre-shared secrets; (2) The protocol is a cross-layer design; (3) Any two appliances in the protocol can establish a secret session key when they do not pre-share any secrets, and it is achieved without much energy usage; (4) The protocol supports key updating and network scalability.
• A Novel Minimum Cost Maximum Power Algorithm for Future Smart Home Energy Management [89] • The proposed "energy management system" aims to decrease the price of consumer's electricity usage by optimizing energy scheduling plan without making any discomfort	• A novel Minimum Cost Maximum Power (MCMP) algorithm to solve the formulated problem of Mixed Integer Linear Programming (MILP). • Price of electricity usage, peak demand reduction, task completion and time of response were analyzed and compared in four different load scenarios.	• Compared to other models: (1) decreases electricity usage and peak demand more efficiently; (2) includes 100% task completion of all home appliances; (3) is less complicated and better performed in real-time situations; (4) is novel and not used before. • Future researches includes energy management in industrial sector.

Title, Category & Objective	The proposed system function and details	Finding or Contribution Limitation & Suggestions
• A Privacy Preserving Communication Protocol for IoT Applications in Smart Homes [90] • The proposed system is for "security system" aimed to present a communication protocol for smart home systems that ensures safety and energy efficiency	• A a Smart Home Systems (SHSs) Protocol for security. It comprises an appliance group, a monitor group, a central controller, and user interfaces. It goes through the following four main phases: (1) key Generation, (2) encryption, (3) Message Authentication Codes (MAC) generation and verification, and (4) decryption. • Safety and privacy-keeping communication protocol applies chaos-based encryption (logistic map) and MACs.	• The first generic architecture design layout for future SHSs. • The system promoted the protocol to achieve better safety and privacy-keeping features with less computations suitable scalability.
• A Security Authorization Scheme for Smart Home IoT Devices [91] • The proposed system is for "security system" to introduce a permission platform for smart home appliances that are connected to unsafe cloud. The instructions are transferred to the user's smart phone for approval.	• The architecture of the proposed security permission platform consists of (1) smart home appliances, (2) cloud platform, (3) smart-phone application. • For the security system and permission platform, the FIDO UAF (Fast Identity Online-Universal Authentication Framework) protocol was selected. The system benefits from open-source IoT cloud, a platform like IoT Kaa, and Several programming languages.	• (1) introducing a platform to create digital ID for both smart appliances and the users, (2) implementing the scheme in an available software and hardware system, (3) designing a cloud federated permission for smart home based on the FIDO permission messages, (4) introducing a security platform in which a user can ask for FIDO permission via cloud platform each time, (5) carrying out tests using Kaa IoT Cloud and assessing delay times.
• Adaptive Monitoring System for E-health Smart Homes [92] • The proposed "healthcare system" aims to introduce a novel adjustable e-health monitoring system for old people.	• An Adaptive Monitoring E-health System, which gathers pertinent data to appraise the user's health condition based on the situations of his daily life activities and is able to learn the activity profile of the user. • Gray Model (GM) is used to predict model (to learn and predict the behavior and level of dependency trends). Markovian model is employed to evaluate the system for generating long term realistic scenarios. • Functional Autonomy Measurement System (SMAF) is pursued to outline the user's activities. Functional abilities adopted were: Activities of Daily Living (ADL), Mobility, Communication, Mental Function and Instrumental Activities of Daily Living (IADL).	• 100% accuracy in high monitoring, 95.8% in medium monitoring, and 91.9% in minimum monitoring. The system diminished energy usage by 48.3%, network traffic by 49.3% and daily activities processing by 54.3%. • Future research include: (1) expanding the number of behavior situations, (2) considering timing logic and break between activities, (3) Including more factors e.g. age, season, location, etc., (4) Incorporating multiple prediction models and assessing effects on the risk and resources, (5) expanding monitoring with vital signs

Title, Category & Objective	The proposed system function and details	Finding or Contribution Limitation & Suggestions
• An Information Provision System to Promote Energy Conservation and Maintain Indoor Comfort in Smart Homes Using Sensed Data by IoT Sensors [93] • The proposed "energy management system" aims to conduct a platform to collect sensing data of environmental parameters (to assess PMV Index) and user's response (comfort desire) to diminish electricity usage while keeping indoor thermal comfort.	• Combines both environmental data and user's response to (1) gather different data from sensors to assess PMV (Predicted Mean Vote) and user's comfort desire, and (2) to choose relevant information from the dataset • PMV Index was assessed by K-means clustering.	Decreased electricity usage by 5.15% and promoted the comfort satisfaction by 42.3%. Qualitative evaluations of indoor comfort enhanced by 16.4%. • Networked sensors gathered data for the system. Future studies: collecting data in a central database operated by a company or the smart city.
• Anonymous Secure Framework in Connected Smart Home Environments [94] • The proposed "security system" aims to provide a secret and safe platform in connected smart homes with less computations and cost.	• An Anonymous Secure Framework (ASF), which creates effective permission and key approval. It provides an unknown and unconnected environment for identity and data of smart devices. Automated Verification of Internet Security Protocol and Application (AVISPA) tool, the BAN-logic and informal analysis are employed.	• (1) less computations and less communication cost; (2) Enhanced safety compared.
• Categories and Functionality of Smart Home Technology for Energy Management [95] • The proposed "energy management system" aims to investigate and recognize various home products on the market and finds out their functions, quality, and their potential for energy savings.	• A coding guide was devised underlying previous work and improved in a repetitive procedure. Qualitative data analysis based on 50 items of the codes.	• Findings indicated choices for saving energy (operational and behavioral) and shifting loads among most product groups. • Future research is to determine the interaction between users and smart home products.
• Context aware Decision Making under Uncertainty for Voice-based Control of Smart Home [96] • The "home management" framework aims to analyze the environment in terms of comfort and independence including voice alarms for immediate actions.	• The decision making and the action is taken with the voice command of a user or when an abnormal situation is discovered by the system. The system is based on a knowledge model and the decision making is processed in uncertainty.	• (1) overall model accuracy level was 85% (min. 71%, max. 100%); (2) Markov Logic Network presented an integrated base involves with logical requirements, uncertainty, and missing data; (3) The system can learn from the data;
• Delivering Home Healthcare through a Cloud-based Smart Home Environment (CoSHE) [97] • The proposed "healthcare system" aims to propose a cloud-centric platform in smart homes to prepare thorough information for health supervisors by monitoring user	• Collecting data via wearable sensors that are not intrusive and also through environmental sensors. A home gateway processes data and sends to a private cloud. This cloud prepares health supervisors with access to real-time information of the user health state. CoSHE consists of four main parts of smart home	• Possible to reach thorough and useful information by unifying context and health-related data and deliver the information to supervisors to monitor the health state of the user. • Limitations: raise energy consumption • Future researches include warning systems to predict the

Title, Category & Objective	The proposed system function and details	Finding or Contribution Limitation & Suggestions
location, activities, mental and emotional state.	environment, wearable devices, private cloud system, and home service robot.	disease and warns health supervisors in risky situations.
• Demand Response Implementation in Smart Households [98] • The proposed "energy management system" aims to propose an optimized-based scheme for energy-saving and cost reduction in smart homes to control the operation of home appliances which their loads can be controlled and are able to store energy (e.g. EV and EWH) considering demand response and the load pricing time.	• Designed for smart houses with photovoltaic (PV) systems combined with Energy Storage System (ESS). Electric vehicle (EV) and electric water heater (EWH) are controllable appliances with storage capability. Effective timing of appliance's energy usage and the demand scheme are scheduled.	• When applying this intelligent optimization-based algorithm, the daily energy price reduced 29.5% - 31.5% comparing to an ordinary rule-based algorithm
• Emulating Home Automation Installations through Component-based Web Technology [99] • The proposed "automation system" of homes aims to introduce a platform based on KNX protocol to imitate, test and validate the home automation environment in a virtual web-based platform before the hardware is installed.	• The proposed component-based web technology platform represents the operation of real devices in a virtual web-based environment based on KNX Network. The virtual devices perform and interact via web environment which permits to make components that create various kinds of data (text, audio, image, video, animation, etc.). COScore structure manages these components. Web services can manipulate the platform and web users' interface can test the behavior regarding this data. • COScore based on CBSE, MDE, web services, HTML5, CSS3 and JavaScript technologies are employed.	• The platform allows testing, validating and analyzing the behavior of home automation devices virtually prior to real installation to minimize risky situations. • Future studies: (1) creating graphical version of the models, (2) applying an MDA approach for the installation, (3) applying security solutions, (4) using another network or protocol apart from KNX.
• Identifying Multiuser Activity with Overlapping Acoustic Data for Mobile Decision Making in Smart Home Environment [100] • The proposed system is for "activity Recognition and movement system" aimed to propose a system to detect activities of users by automatically identifying the various and overlapping sounds in a real-time context in smart homes using acoustic-based LED lighting.	• Sounds of user's activities are collected and detected by audio sensors in a real-time manner. The sounds are classified for activity recognition.	• Higher accuracy rate than existing classification algorithms. Future studies: Improvement of smart recognition systems with further experiments in real-time manner.
• ImAtHome-making Trigger-action Programming Easy and Fun [101]	• For every home, the application makes a database on the cloud with different	• Majority of the respondents particularly young people found ImAtHome user-

Title, Category & Objective	The proposed system function and details	Finding or Contribution Limitation & Suggestions
• The proposed "automation and home management systems" aims to propose a framework for smart home users to control smart appliances by setting certain commands for operations without having any expertise.	groups and sub-groups of various rooms and devices. It can find, add or remove rooms and devices, access their data, and control or command (even vocally) different functions.	friendly. • Future work includes: applying more users and tasks with different scenarios based on different application platforms e.g. MicroApp.
• IoTtalk-RC: Sensors as Universal Remote Control for Aftermarket Home Appliances [102] • The proposed "automation and home management systems" aims to introduce a software-based platform for distance-control of smart home devices	• The proposed system is IoTtalk Remote Control (IoTtalk-RC), which uses sensors (e.g. temperature & humidity Sensors) to control one or more tasks of smart home devices (e.g. electric fans). An infrared device for receiving (IrR-D) and one for transmitting (IrT-D) are connected to IoTtalk server which can be installed in a cloud or Wi-Fi AP. The Smart device is connected to the IoTtalk server via wireless system. Computations and analytic analysis are employed.	• (1) Short delay time even when IoTtalk server is far at a virtual machine of a cloud; (2) The disorganized series of sending due to the delay time is insignificant; (3) efficient platform; (4) IoTtalk is a solution to replace the usual infrared remote controller of smart devices needs new programming from the developers.

Table 8.
Reviewed papers related to smart homes.

Author details

Reyhaneh Karimi[1*], Leila Farahzadi[2], Samad M.E. Sepasgozar[3], Sharifeh Sargolzaei[4], Sanee M. Ebrahimzadeh Sepasgozar[5], Mohsen Zareian[6] and Akram Nasrolahi[2]

1 The University of Science and Culture, Tehran, Iran

2 Islamic Azad University, Tehran, Iran

3 The University of New South Wales, Sydney, NSW, 2052, Australia

4 University of Zabol, Zabol, Iran

5 The Knowledge Enterprise Payesh Tajhiz Arad, Gorgan, 1734533998, Iran

6 Chalmers University of Technology, Göteborg, 412 96, Sweden

*Address all correspondence to: reyhanehkarimi@rocketmail.com

References

[1] S. M. Sepasgozar, S. Hawken, S. Sargolzaei, and M. Foroozanfa, "Implementing citizen centric technology in developing smart cities: A model for predicting the acceptance of urban technologies," *Technol. Forecast. Soc. Change*, no. 142, pp. 105–116, 2019.

[2] C. Harrison and I. A. Donnelly, "A Theory of Smart Cities," in *Proceedings of the 55th Annual Meeting of the ISSS*, 2011, pp. 1–15.

[3] R. Hall, "The vision of a smart city International Life Extension Technology Workshop Paris," Fr. Sept., vol. 28, p. 2000, 2000.

[4] S. M. Sepasgozar, F. Tahmasebinia, and S. Shirowzhan, *Infrastructure Management and Construction*. 2020.

[5] T. Nam and T. A. Pardo, "Conceptualizing smart city with dimensions of technology, people, and institutions," in *Proceedings of the 12th Annual International Digital Government Research Conference: Digital Government Innovation in Challenging Times*, 2011, pp. 282–291.

[6] T. Bakıcı, E. Almirall, and J. Wareham, "A smart city initiative: the case of Barcelona," J. Knowl. Econ., vol. 4, no. 2, pp. 135–148, 2013.

[7] L.-G. Cretu, "Smart cities design using event-driven paradigm and semantic web," Inform. Econ., vol. 16, no. 4, p. 57, 2012.

[8] R. Giffinger, C. Fertner, H. Kramar, R. Kalasek, N. Pichler-Milanović, and E. Meijers, "Smart cities: Ranking of european medium-sized cities. vienna, austria: Centre of regional science (srf), vienna university of technology," *www. smart-cities. eu/download/smart cities Final report. pdf*, 2007.

[9] S. Shirowzhan, W. Tan, and S. M. Sepasgozar, "Digital Twin and CyberGIS for Improving Connectivity and Measuring the Impact of Infrastructure Construction Planning in Smart Cities," ISPRS Int. J. Geo-Information, vol. 9, no. 4, p. 240, 2020.

[10] S. Shirowzhan, S. Lim, J. Trinder, H. Li, and S. M. Sepasgozar, "Data mining for recognition of spatial distribution patterns of building heights using airborne lidar data," *Adv. Eng. Informatics*, no. 43, p. 101033, 2020.

[11] S. M. E. Sepasgozar, H. Li, S. Shirowzhan, and V. W. Y. Tam, "Methods for monitoring construction off-road vehicle emissions: a critical review for identifying deficiencies and directions," Environ. Sci. Pollut. Res., vol. 26, no. 16, pp. 15779–15794, 2019, doi: 10.1007/s11356-019-05003-6.

[12] V. Albino, U. Berardi, and R. M. Dangelico, "Smart cities: Definitions, dimensions, performance, and initiatives," J. Urban Technol., vol. 22, no. 1, pp. 3–21, 2015.

[13] P. Lombardi, S. Giordano, H. Farouh, and W. Yousef, "Modelling the smart city performance Innovation," *Eur. J. Soc. Sci. Res.*, vol. 25, no. 2, pp. 137–149, 2012.

[14] R. Giffinger and H. Gudrun, "Smart cities ranking: an effective instrument for the positioning of the cities?," ACE Archit. city Environ., vol. 4, no. 12, pp. 7–26, 2010.

[15] J. V. Winters, "Why are Smart Cities Growing? Who Moves and Who Stays," J. Reg. Sci., vol. 51, no. 2, pp. 253–270, 2011.

[16] M. B. Hoy, "Smart Buildings: An Introduction to the Library of the Future," Med. Ref. Serv. Q., vol. 35, no. 3, pp. 326–331, 2016, doi: 10.1080/02763869.2016.1189787.

[17] J. A. Powell, "Intelligent design teams design intelligent buildings," Habitat Int., vol. 14, no. 2 & 3, pp. 83–94, 1990.

[18] R. Everett, "The building colleges for the future program: Delivering a green and intelligent building agenda," New Rev. Inf. Netw., vol. 14, no. 1, pp. 3–20, 2008.

[19] K. P. Wacks, "Best-Practices Guide for Evaluating Intelligent Building Technologies," 2002.

[20] O. Frank, "Intelligent building concept: the challenges for building practitioners in the 21st century," J. Assoc. Archit. Educ. Niger. (AARCHES J), vol. 6, no. 3, pp. 107–113, 2007.

[21] J. K. W. Wong, H. Li, and S. W. Wang, "Intelligent building research: a review," Autom. Constr., vol. 14, no. 1, pp. 143–159, 2005.

[22] R. Panchalingam and K. C. Chan, "A state-of-the-art review on artificial intelligence for Smart Buildings," Intell. Build. Int., vol. 0, no. 0, pp. 1–24, 2019, doi: 10.1080/17508975.2019.1613219.

[23] a. H. Buckman, M. Mayfield, and S. B. M. Beck, "What is a smart building?," Smart Sustain. Built Environ., vol. 3, no. 2, pp. 92–109, 2014, doi: 10.1108/SASBE-01-2014-0003.

[24] CABA, *Bright Green Buildings: Convergence of Green and Intelligent Buildings*. Continental Automated Buildings Association (CABA), 2008.

[25] Y. B. Agarwal, R. Balaji, J. Gupta, M. W. Lyles, and T. Weng, "Occupancy-driven energy management for smart building automation," BuildSys, IEEE, 2010.

[26] M. V. Moreno, M. A. Zamora, and A. F. Skarmeta, "User-centric smart buildings for energy sustainable smart cities," Trans. Emerg. Telecommun. Technol., vol. 25, no. 1, pp. 41–55, 2014.

[27] J. Sinopoli, *Smart Buildings Systems for Architects, Owners and Builders*. 2010.

[28] Owen King, "Smart Working: Smart Buildings and Future of Work," 2016.

[29] Jennifer King and Christopher Perry, "Smart Buildings: Using Smart Technology to Save Energy in Existing Buildings," *Am. Counc. an Energy-Efficient Econ.*, no. February, pp. 1–46, 2017.

[30] L. Linder, D. Vionnet, J. P. Bacher, and J. Hennebert, "Big Building Data-a Big Data Platform for Smart Buildings," Energy Procedia, vol. 122, pp. 589–594, 2017, doi: 10.1016/j.egypro.2017.07.354.

[31] M. R. Bashir and A. Q. Gill, "IoT enabled smart buildings: A systematic review," *2017 Intell. Syst. Conf. IntelliSys 2017*, vol. 2018-Janua, no. September, pp. 151–159, 2018, doi: 10.1109/IntelliSys.2017.8324283.

[32] H. Said, *Smart Buildings and Internet of Things (Iot) Impact on Electrical*, no. January. 2018, pp. 1–104.

[33] J. K. Wong and H. Li, "Application of the Analytic Hierarchy Process (AHP) in Multi-Criteria Analysis of the Selection of Intelligent Building Systems," Build. Environ., vol. 43, no. 108–125, 2008.

[34] G. I. Fântână and S. A. Oae, "Evolution of Smart Buildings," in *International Conference on Environment, Energy, Ecosystems and Development*, 2013, pp. 223–225.

[35] Control Solution, "Evolution of smart buildings," *Control Solution Inc.*, 2020. [Online]. Available: http://controlyourbuilding.com/the-evolution-of-smart-buildings. [Accessed: 20-Mar-2020].

[36] Siemens, "The future of smart building," 2018.

[37] A. Ghaffarianhoseini *et al.*, "What is an intelligent building? Analysis of recent interpretations from an international perspective," Archit. Sci. Rev., vol. 59, no. 5, pp. 338–357, 2015, doi: 10.1080/00038628.2015.1079164.

[38] D. J. Clements-Croome, *Intelligent Buildings: Design, Management and Operation*, 2nd ed. London: ICT Publishing, 2013.

[39] F. Tahmasebinia *et al.*, "Criteria development for sustainable construction manufacturing in Construction Industry 4.0," *Constr. Innov.*, 2020.

[40] A. Verma, S. Prakash, V. Srivastava, A. Kumar, and S. C. Mukhopadhyay, "Sensing, Controlling, and IoT Infrastructure in Smart Building: A Review," IEEE Sens. J., vol. 19, no. 20, pp. 9036–9046, 2019, doi: 10.1109/JSEN.2019.2922409.

[41] S. Shirowzhan, S. M. Sepasgozar, D. J. Edwards, H. Li, and C. Wang, "BIM compatibility and its differentiation with interoperability challenges as an innovation factor," *Autom. Constr.*, no. 112, p. 103086, 2020.

[42] I. V. Lokshina, M. Greguš, and W. L. Thomas, "Application of integrated building information modeling, iot and Blockchain Technologies in System Design of a Smart Building," Procedia Comput. Sci., vol. 160, pp. 497–502, 2019, doi: 10.1016/j.procs.2019.11.058.

[43] S. Clarke, "The Future Of Smart Buildings," 2018.

[44] E. I. Batov, "The distinctive features of 'smart' buildings," *Procedia Eng.*, vol. 111, no. TFoCE, pp. 103–107, 2015, doi: 10.1016/j.proeng.2015.07.061.

[45] M. Jia, A. Komeily, Y. Wang, and R. S. Srinivasan, "Adopting Internet of Things for the development of smart buildings: A review of enabling technologies and applications," *Autom. Constr.*, vol. 101, no. January, pp. 111–126, 2019, doi: 10.1016/j.autcon.2019.01.023.

[46] J. Aamidor, "The three types of smart building systems: the way forward for practitioners and vendors," *medium*, 2016. .

[47] D. Djenouri, R. Laidi, Y. Djenouri, and I. Balasingham, "Machine learning for smart building applications: Review and taxonomy," ACM Comput. Surv., vol. 52, no. 2, 2019, doi: 10.1145/3311950.

[48] X. Zhang, M. Pipattanasomporn, T. Chen, and S. Rahman, "An IoT-Based Thermal Model Learning Framework for Smart Buildings," IEEE Internet Things J., vol. 7, no. 1, pp. 518–527, 2020, doi: 10.1109/JIOT.2019.2951106.

[49] J. R. Robles and T. Kim, "Applications, Systems and Methods in Smart Home Technology: A Review," *Int. J. Adv. Sci. Technol.*, vol. 15, no. November, pp. 37–48, 2010.

[50] M. Alaa, a. a. Zaidan, B. B. Zaidan, M. Talal, and M. L. M. Kiah, "A review of smart home applications based on Internet of Things," J. Netw. Comput. Appl., vol. 97, pp. 48–65, 2017, doi: 10.1016/j.jnca.2017.08.017.

[51] S. Sepasgozar *et al.*, "A Systematic Content Review of Artificial Intelligence and the Internet of Things Applications in Smart Home," Appl. Sci., vol. 10, no. 9, p. 3074, 2020.

[52] M. Domb, "Smart Home Systems Based on Internet of Things," in *IntechOpen*, vol. i, no. tourism, 2016, p. 13.

[53] G. Lobaccaro, S. Carlucci, and E. Löfström, "A review of systems and technologies for smart homes and smart grids," Energies, vol. 9, no. 5, pp. 1–33, 2016, doi: 10.3390/en9050348.

[54] J. El Hachem, V. Chiprianov, M. A. Babar, T. Al Khalil, and P. Aniorte, "Modeling, analyzing and predicting security cascading attacks in smart buildings systems-of-systems," J. Syst. Softw., vol. 162, p. 110484, 2020, doi: 10.1016/j.jss.2019.110484.

[55] K. Srivastava, "Prediction Model for Personal Thermal Comfort for Naturally Ventilated Smart Buildings," in *Proceedings of ICETIT*, 2019, pp. 117–127, doi: https://doi.org/10.1007/978-3-030-30577-2_10.

[56] H. Sattar *et al.*, "An intelligent and smart environment monitoring system for healthcare," *Appl. Sci.*, vol. 9, no. 19, 2019, doi: 10.3390/app9194172.

[57] H. Park, S. Hwang, M. Won, and T. Park, "Activity-aware sensor cycling for human activity monitoring in smart homes," IEEE Commun. Lett., vol. 21, no. 4, pp. 757–760, 2016.

[58] I. Bisio, A. Delfino, F. Lavagetto, and A. Sciarrone, "Enabling IoT for in-home rehabilitation: Accelerometer signals classification methods for activity and movement recognition," IEEE Internet Things J., vol. 4, no. 1, pp. 135–146, 2016.

[59] W. Yang, X. Jing, and H. Huang, "An Intelligent Human Behavior-Based Reasoning Model for Service Prediction in Smart Home," IEEE Access, vol. 6, pp. 68535–68544, 2018.

[60] A. Alaiad and L. Zhou, "Patients' adoption of WSN-based smart home healthcare systems: An integrated model of facilitators and barriers," IEEE Trans. Prof. Commun., vol. 60, no. 1, pp. 4–23, 2017.

[61] C. Wilson, T. Hargreaves, and R. Hauxwell-baldwin, "Benefits and risks of smart home technologies," vol. 103, no. January, pp. 72–83, 2017, doi: 10.1016/j.enpol.2016.12.047.

[62] B. Lee, O. Kwon, I. Lee, and J. Kim, "Companionship with Smart Home Devices: The Impact of Social Connectedness and Interaction Types on Perceived Social Support and Companionship in Smart Homes," *Comput. Human Behav.*, 2017, doi: 10.1016/j.chb.2017.06.031.

[63] E. Park, Y. Cho, J. Han, S. J. Kwon, A. Iot, and S. Home, "Comprehensive Approaches to User Acceptance of Internet of Things in a Smart Home Environment," vol. 4, no. 6, pp. 2342–2350, 2017.

[64] Y. Strengers and L. Nicholls, "Convenience and energy consumption in the smart home of the future-industry visions from Australia and beyond," *Chem. Phys. Lett.*, 2017, doi: 10.1016/j.erss.2017.02.008.

[65] P. Verma and S. K. Sood, "Fog assisted-IoT enabled patient health monitoring in smart homes," IEEE Internet Things J., vol. 5, no. 3, pp. 1789–1796, 2018.

[66] J. Dahmen, D. J. Cook, X. Wang, and W. Honglei, "Smart secure homes: a survey of smart home technologies that sense, assess, and respond to security threats," J. Reliab. Intell. Environ., vol. 3, no. 2, pp. 83–98, 2017.

[67] H. Xu, L. König, D. Cáliz, and H. Schmeck, "A generic user interface for energy management in smart homes," Energy Informatics, vol. 1, no. 1, p. 55, 2018.

[68] Y. Li *et al.*, "A Health Gateway for Mobile Monitoring in Nursing Home," Wirel. Pers. Commun., vol. 102, no. 2, pp. 1573–1587, 2018.

[69] L.-P. Hung, Y.-H. Chao, and C.-L. Chen, "A Hybrid Key Item Locating Method to Assist Elderly Daily Life Using Internet of Things," Mob. Networks Appl., vol. 24, no. 3, pp. 786–795, 2019.

[70] W. G. de Oliveira Júnior, J. M. de Oliveira, R. Munoz, and V. H. C. de Albuquerque, "A proposal for Internet of Smart Home Things based on BCI system to aid patients with amyotrophic lateral sclerosis," *Neural Comput.* Appl., pp. 1–11, 2018.

[71] M. Amiribesheli and H. Bouchachia, "A tailored smart home for dementia care," J. Ambient Intell. Humaniz. Comput., vol. 9, no. 6, pp. 1755–1782, 2018.

[72] P. Carnemolla, "Ageing in place and the internet of things–how smart home technologies, the built environment and caregiving intersect," Vis. Eng., vol. 6, no. 1, p. 7, 2018.

[73] D. R. Dhatchayeny, W. A. Cahyadi, and Y.-H. Chung, "An Assistive VLC Technology for Smart Home Devices Using EOG," Wirel. Pers. Commun., vol. 98, no. 1, pp. 81–89, 2018.

[74] K. Naik and S. Patel, "An open source smart home management system based on IOT," Wirel. Networks, pp. 1–7, 2018.

[75] D. Popa, F. Pop, C. Serbanescu, and A. Castiglione, "Deep learning model for home automation and energy reduction in a smart home environment platform," Neural Comput. Appl., vol. 31, no. 5, pp. 1317–1337, 2019.

[76] D. Behrens, T. Schoormann, S. Bräuer, and R. Knackstedt, "Empowering the selection of demand response methods in smart homes: development of a decision support framework," Energy Informatics, vol. 1, no. 1, p. 53, 2018.

[77] G. Mokhtari, S. Aminikhanghahi, Q. Zhang, and D. J. Cook, "Fall detection in smart home environments using UWB sensors and unsupervised change detection," J. Reliab. Intell. Environ., vol. 4, no. 3, pp. 131–139, 2018.

[78] K. W. G. Cowdrey and R. Malekian, "Home automation-an IoT based system to open security gates using number plate recognition and artificial neural networks," Multimed. Tools Appl., vol. 77, no. 16, pp. 20325–20354, 2018.

[79] H. Jo and Y. I. Yoon, "Intelligent smart home energy efficiency model using artificial TensorFlow engine," Human-centric Comput. Inf. Sci., vol. 8, no. 1, p. 9, 2018.

[80] G. Chen, A. Wang, S. Zhao, L. Liu, and C.-Y. Chang, "Latent feature learning for activity recognition using simple sensors in smart homes," Multimed. Tools Appl., vol. 77, no. 12, pp. 15201–15219, 2018.

[81] S. N. Makhadmeh, A. T. Khader, M. A. Al-Betar, and S. Naim, "Multi-objective power scheduling problem in smart homes using grey wolf optimiser," J. Ambient Intell. Humaniz. Comput., pp. 1–25, 2018.

[82] B. Hussain and Q. U. L. Hasan, "An Innovative Heuristic Algorithm for IoT-Enabled Smart Homes for Developing Countries," IEEE Access, vol. 6, pp. 15550–15575, 2018, doi: 10.1109/ACCESS.2018.2809778.

[83] W. Li, G. S. Member, T. Logenthiran, S. Member, and A. S. Homes, "A Novel Smart Energy Theft System (SETS) for IoT-Based Smart Home," IEEE Internet Things J., vol. 6, no. 3, pp. 5531–5539, 2019, doi: 10.1109/JIOT.2019.2903281.

[84] A. M. Zungeru, S. Member, and J. Gaboitaolelwe, "A Secured Smart Home Switching System based on Wireless Communications and Self-Energy Harvesting," IEEE Access, vol. 7, pp. 25063–25085, 2019, doi: 10.1109/ACCESS.2019.2900305.

[85] R. Teng and T. Yamazaki, "Load Profile-Based Coordination of

Appliances in a Smart Home," IEEE Trans. Consum. Electron., vol. 65, no. 1, pp. 38–46, 2019, doi: 10.1109/TCE.2018.2885874.

[86] L. Yu, T. Jiang, and Y. Zou, "Online Energy Management for a Sustainable Smart Home With an HVAC Load and Random Occupancy," IEEE Trans. Smart Grid, vol. 10, no. 2, pp. 1646–1659, 2019, doi: 10.1109/TSG.2017.2775209.

[87] M. Rawashdeh, M. Gh, A. Zamil, S. Samarah, M. S. Hossain, and G. Muhammad, "A knowledge-driven approach for activity recognition in smart homes based on activity profiling," 2017, doi: 10.1016/j.future.2017.10.031.

[88] Y. Zhang, Y. Xiang, X. Huang, and X. Chen, "A matrix-based cross-layer key establishment protocol for smart homes," Inf. Sci. (Ny)., vol. 429, no. 2018, pp. 390–405, 2020, doi: 10.1016/j.ins.2017.11.039.

[89] A. Singaravelan and M. Kowsalya, "A Novel Minimum Cost Maximum Power Algorithm for Future Smart Home Energy Management . School of Electrical Engineering," *J. Adv. Res.*, 2017, doi: 10.1016/j.jare.2017.10.001.

[90] T. Song, R. Li, B. Mei, J. Yu, X. Xing, and X. Cheng, "A Privacy Preserving Communication Protocol for IoT Applications in Smart Homes," vol. 4, no. 6, pp. 1844–1852, 2017.

[91] B. Chifor, I. Bica, V. Patriciu, and F. Pop, "A security authorization scheme for smart home Internet of Things devices," *Futur. Gener. Comput. Syst.*, 2017, doi: 10.1016/j.future.2017.05.048.

[92] H. Mshali, T. Lemlouma, and D. Magoni, "Adaptive Monitoring System for e-Health Smart Homes," *Pervasive Mob. Comput.*, 2017, doi: 10.1016/j.pmcj.2017.11.001.

[93] K. Matsui, "An information provision system to promote energy conservation and maintain indoor comfort in smart homes using sensed data by IoT sensors," *Futur. Gener. Comput. Syst.*, 2017, doi: 10.1016/j.future.2017.10.043.

[94] J. Iinatti, S. Member, and P. H. Ha, "Anonymous secure framework in connected smart home environments," IEEE Trans. Inf. Forensics Secur., vol. 12, no. 4, pp. 968–979, 2017, doi: 10.1109/TIFS.2016.2647225.

[95] R. Ford *et al.*, "Categories and functionality of smart home technology for energy management," *Build. Environ.*, 2017, doi: 10.1016/j.buildenv.2017.07.020.

[96] P. Chahuara, F. Portet, and M. Vacher, "Context-aware decision making under uncertainty for voice-based control of smart home," vol. 75, pp. 63–79, 2017, doi: 10.1016/j.eswa.2017.01.014.

[97] M. Pham, Y. Mengistu, H. Do, and W. Sheng, "Delivering home healthcare through a Cloud-based Smart Home Environment (CoSHE)," 2017, doi: 10.1016/j.future.2017.10.040.

[98] M. Ali, F. Ghazvini, J. Soares, O. Abrishambaf, R. Castro, and Z. Vale, "Demand response implementation in smart households," *Energy Build.*, 2017, doi: 10.1016/j.enbuild.2017.03.020.

[99] J. A. Asensio, J. Criado, N. Padilla, and L. Iribarne, "Emulating Home Automation Installations through Component-based Web Technology," *Futur. Gener. Comput. Syst.*, 2017, doi: 10.1016/j.future.2017.09.062.

[100] J. S. Lee, S. Choi, and O. Kwon, "Identifying multiuser activity with overlapping acoustic data for mobile decision making in smart home environment," *Expert Syst. Appl.*, 2017, doi: 10.1016/j.eswa.2017.03.062.

[101] D. Fogli, M. Peroni, and C. Stefini, "ImAtHome-making trigger-action programming easy and fun," J. Vis. Lang. Comput., vol. 42, pp. 60–75, 2017, doi: 10.1016/j.jvlc.2017.08.003.

[102] Y. Lin, Y. Lin, C. Hsiao, and Y. Wang, "IoTtalk-RC-sensors as universal remote control for aftermarket home appliances," vol. 4, no. 4, pp. 1104–1112, 2017.